Measurement Uncertainty:
A Reintroduction

Measurement Uncertainty: A Reintroduction

Antonio Possolo & Juris Meija

sistema interamericano de metrologia — sim

Measurement Uncertainty: A Reintroduction, 2nd edition
Published by SISTEMA INTERAMERICANO DE METROLOGIA — SIM
Montevideo, Uruguay: 2022

ISBN 978-0-660-42866-6 (PRINT)
ISBN 978-0-660-42865-9 (PDF)
DOI 10.4224/1tqz-b038

Cover design by Neeko Paluzzi, including original, personal artwork © 2016 Juris Meija. Front matter, body, and end matter composed with LaTeX as implemented in the 2021 TeX Live distribution from the TeX Users Group, using the `tufte-book` class (© 2007-2015 Bil Kleb, Bill Wood, and Kevin Godby), with Palatino, Helvetica, and Bera Mono typefaces.

It is the uncertainty that charms one.
A mist makes things wonderful.
— Oscar Wilde (1890, *The Picture of Dorian Gray*)

Contents

Preface

Our original aim was to write an introduction to the evaluation and expression of measurement uncertainty as accessible and succinct as Stephanie Bell's little jewel of *A Beginner's Guide to Uncertainty of Measurement* [Bell, 1999], only showing a greater variety of examples to illustrate how measurement science has grown and widened in scope in the course of the intervening twenty years.

The recent, very welcome *Introduction to Measurement Uncertainty* that Blair Hall and Rod White have made available to the community [Hall and White, 2018], occupies a middle ground in terms of complexity. It presents two realistic examples in considerable detail (using a ruler, and calibrating a thermometer), and it excels in typographical design, from which we have obviously drawn inspiration.

Our account turned out far more ambitious and challenging then either of these two that motivated us and that we had intended to emulate and expand upon, "only a little." In fact we assume that the reader already is familiar with either, or preferably with both of them. For this reason, we have characterized our contribution as a reintroduction to measurement uncertainty.

We take an eclectic and inclusive view of measurement, recognizing its vital and pervasive role in science and technology, also in the arts. Since the interests of individual readers may be more narrowly focused, we have organized our narrative so that a reader who is primarily interested in weighing may skip the discussion of counting, and similarly for all the other sections.

Even subsections within the same section can, in most cases, be read independently of one another: for example, to learn how to compare two measurement methods, while remaining unconcerned with how to compare a measured value with a corresponding certified value.

The price to be paid for such flexibility is the amount of internal cross-referencing, either via page numbers (in the print edition), or via hyperlinks (in the online version).

While some of our examples are very simple and likely to appeal to a broad audience (measuring the volume of a storage tank, or surveying a plot of land), others may interest only a more narrowly specialized sector of the readership (measuring abortion rates, or calibrating a resistor using a Wheatstone bridge).

Some applications may appear, at first blush, to be narrowly focused (measuring the Hubble-Lemaître constant), but in fact illustrate techniques that are widely applicable. Still others are fairly complex, yet are likely to draw the attention of many readers (calibrating a GC-MS system, or averaging models for a flu epidemic).

The predominant approach to measurement uncertainty involves probabilistic concepts and requires the application of statistical methods. We have chosen not to hide the attending difficulties, and strive to explain the models we use, and the calculations necessary to apply them, in fair detail, providing computer codes to carry them out.

These technicalities, no matter how clearly one may be able to explain them, inevitably will be challenging obstacles for many readers. Two appendices, one on probability, the other on statistics, may help motivated readers familiarize themselves with concepts and methods sufficiently to overcome such obstacles, yet they demand considerable commitment from the reader.

We offer supplementary material online in a companion web site, including datasets, computer codes, and problems and exercises likely to be helpful for instruction based on the book.

We apply a wide range of statistical models and methods, some from the classical school, others of a Bayesian flavor, especially when it is advantageous to incorporate preexisting knowledge about a measurand, or to nudge a procedure in a particular direction. We eschew the rigidity entailed by ideological purity, and willingly employ any tool or approach that seems best suited for the task at hand.

The key resolution we made was to approach each problem with flexibility, being deferential to the data and attentive to the purpose of the inquiry: to select models and employ data reduction techniques that are verifiably adequate for the data in hand; to give each problem a custom solution tailored for the purpose that the results are intended to serve; all along heeding Lincoln Moses's advice that "You have to have a good data-side manner."

Acknowledgments & Disclaimers

The authors are honored by the support the SIM Council has chosen to grant to this contribution by approving its publication under the SIM banner: Claire Saundry (NIST), *President*; Hector Laiz (INTI), *Former President*; J. Salvador Echeverria Villagomez (CENAM), *Technical Committee Chair*; Sally Bruce (NIST), QSTF *Chair*; Javier Arias (CENAMEP), *Project Coordinator*; Rodrigo Costa-Felix (INMETRO), *Professional Development Coordinator*; Edwin Arvey Cristancho-Pinilla (INM), ANDIMET *Coordinator*; Pedro Ibarra (INN), SURAMET *Coordinator*; Claudia Estrada (CIM), CAMET *Coordinator*; I-Ronn Audain (SKNBS), CARIMET *Coordinator*; and Victor Lizardi (CENAM), NORAMET *Coordinator*.

The example involving a Wheatstone bridge is patterned after a laboratory experiment of Physics 321 as taught by Dan McCammon and Vincent Liu at the University of Wisconsin-Madison, in the fall of 2016.

John Sieber, Tom Vetter, and Adam Pintar, all from NIST, shared the observations of fluorescence intensity used to illustrate a permutation test for homogeneity of a reference material. George D. Quinn (Material Measurement Laboratory, NIST), provided a set of determinations of rupture strength of alumina coupons. Laura Wood and Catherine Rimmer (Material Measurement Laboratory, NIST) gave us permission to use determinations of the mass fraction of arsenic in kudzu.

The readings of temperature of a thermal bath (Page 34) were generously shared by Victor Eduardo Herrera Diaz (*Centro de Metrología del Ejército Ecuatoriano*, CMME, Quito, Ecuador) during an international workshop held at the *Laboratorio Tecnológico del Uruguay* (LATU, Montevideo, Uruguay) in March, 2013. The example of linear interpolation for the pH of beer (Page 86) was motivated by conversations with Walter Miller and Christina Cecelski, both from NIST. Leticia Pibida, also from NIST, provided information about alpha particle spectrometry. Olha Bodnar (Örebro University, Sweden) suggested improvements for the example about Counting Radon Atoms (Page 46).

Several colleagues from SIM countries provided valuable comments, and pointed out errors. We are particularly grateful to Hugo Gasca Aragón (CENAM, Mexico), Bryan Calderón (LACOMET, Costa Rica), Silvina Forastieri (INTI, Argentina), Hari Iyer (NIST), Dianne Lalla-Rodrigues (ABBS, Antigua and Barbuda), Wilson Naula (INEN, Ecuador), Claudia Santo (SIM), Barry Wood, Carlos Sanchez, and Brad Methven (NRC, Canada), for their generous contributions and perceptive insights.

Mary Gregg and Jack Prothero (both from NIST), and Kesten Bozinovic (Georgetown University, Washington, DC), reviewed the contents throughly, detected many errors and provided copious comments and suggestions that induced substantial improvements to the contents.

The *Statistics* Appendix (Page 177) suggests that statistics is an art that one learns from master artisans. Antonio Possolo was fortunate to have apprenticed with John Hartigan (Yale), Frank Anscombe (Yale), and John Tukey (Princeton).

Measurement

Our ancestors were shepherds that counted sheep, surveyors that sized agricultural land, traders that weighed gold pieces, time-keepers that relied on sundials, merchants that graded silk according to its fineness, and healers that assigned medicinal plants to categories reflecting their powers (*cf.* Todd [1990]).

In ancient Egypt, measurement was considered important even in the afterlife: Anubis (god of death) leads the scribe Hunefer to judgement, where his heart is weighed against the Feather of Truth. Thoth (god of writing) records the result, while Ammit, Devourer of the Dead, awaits the verdict.
— *Book of the Dead* (1275 BCE) British Museum (EA 9901,3)

Counting, surveying, weighing, timing, ranking, and classifying all serve to assign a value to a property (*measurand*) of an object of interest, and all are instances of measurement provided they satisfy these requirements: (i) the assignment of value is based on comparison with a standard that is recognized as a common reference by the community of producers and users of the measurement result; (ii) the measured value is qualified with an evaluation of measurement uncertainty whose practical meaning is well understood and agreed upon; (iii) the measurement result (measured value together with its associated measurement uncertainty) is used to inform an action or decision [White, 2011] [Possolo, 2018].

Recognizing and quantifying the uncertainty that invariably clouds our knowledge of the world is a hallmark of science. It informs actions and decisions in all fields of the human endeavor: protecting against incoming storms, planning crops, responding to epidemics, or managing industrial inventories. Measurement uncertainty is an integral part of every measurement result, characterizing its quality.

A measured value is an estimate of the true value of a property, which may be quantitative or qualitative. Counting, surveying, weighing, and timing all produce estimates of quantitative measurands. Ranking applies to qualities whose values can meaningfully be ordered from lowest to highest, or weakest to strongest (for example, the Mohs hardness of a mineral, or the spiciness of a curry). Classification (or identification) assigns objects to categories that are either identical or different, but that cannot otherwise be ordered or quantified (for example, the identity of the nucleobase at a particular location of a DNA strand, or the gender of an athlete).

Measurement Uncertainty

Measurement uncertainty is the doubt about the true value of the measurand that remains after making a measurement [Possolo, 2015]. The corresponding margin of doubt is characterized by its width (size of the uncertainty) and by its depth (severity of the uncertainty): the wider this margin, the larger the range of values of the measurand that are consistent with the measured value; the deeper this margin, the smaller the confidence that the true value of the measurand lies within that margin [Bell, 1999].

Truth lies hidden in a castle's keep, surrounded by uncertainty, which is represented by the moat. The width of the moat portrays the margin of doubt, and its depth illustrates the severity of the doubt [Bell, 1999] (Almourol Castle, Portugal — Wikimedia Commons, Daniel Feliciano, 2003).

There is no science without measurements, no quality without testing, and no global commerce without standards. Since no measurement is perfect, evaluating measurement uncertainty and taking it into account are prerequisites for using measurement results.

Uncertainty often originates not only from imperfections in measurement, but also from the natural variability of the true values of the properties we seek to measure. For example, the exact amount of aspirin may vary slightly among nominally identical pills, and the actual volume of dishwashing liquid in bottles supposed to contain the same nominal volume often varies enough to be perceptible to the naked eye.

The speed of light in vacuum has exactly one true value that is invariant in time and space, according to the prevailing view of the universe. But the true value of the atomic weight of oxygen varies significantly across USA river waters, reflecting the spatial variability of the amount fractions of its different isotopes [Kendall and Coplen, 2001].

In addition to imperfect measurements or natural variability of the true values of measurands, it is common for there to be ambiguity, or incomplete specification, of the very definition of what we are trying to measure. The following three examples describe cases where such ambiguity was an important source of uncertainty.

In January, 2015, the U.S. Supreme Court decided a case concerning the meaning of the term "molecular weight" as it had been used in a patent filed by Teva. The Court considered that "the term might refer to the weight of the most numerous molecule, it might refer to weight as calculated by the average weight of all molecules, or it might refer to weight as calculated by an average in which heavier molecules count for more."[1]

Driving under the influence (DUI) court cases rely on measurements made to determine whether alcohol concentration exceeded 0.08 g per 100 mL of blood, or 0.08 g per 210 L of breath. Typically, the prosecution has to demonstrate that the alcohol concentration indeed exceeded the 0.08 level beyond reasonable doubt, which is often taken to mean 99 % confidence.

[1] *Teva Pharmaceuticals USA, Inc.* v. *Sandoz, Inc.* 574 U. S. 318 (2015), 2015

) Case No. C076949 and 9Y6231062
)
) ORDER SUPPRESSING DEFENDANT'S
) BREATH-ALCOHOL MEASUREMENTS IN
) THE ABSENCE OF A MEASUREMENT
) FOR UNCERTAINTY

Measurement uncertainty is crucial to determining whether laws are broken (excerpt from a 2010 King County District Court ruling, Washington, USA).

Besides the sizable measurement uncertainty, which is in large part attributable to calibration uncertainty,[2] [3] the factors affecting the outcome of breath tests include body temperature, blood makeup (hematocrit, the volume fraction of red blood cells in the blood), and the manner of breathing.

[2] S. Cowley and J. Silver-Greenberg. These Machines Can Put You in Jail. Don't Trust Them. *The New York Times*, November 3, 2019. Business Section

[3] J. Silver-Greenberg and S. Cowley. 5 Reasons to Question Breath Tests. *The New York Times*, November 3, 2019. Business Section

Moreover, uncertainty can surround many other aspects of the measurement: some parts of the body will have higher blood-alcohol concentration than others, with the alcohol levels in arterial and venous blood possibly differing by as much as a factor of two [Simpson, 1987].

Even the very definition of alcohol, surprisingly, can include not only ethanol but also other low molecular weight alcohols such as methanol or isopropanol.

Defining gender, in particular of athletes participating in sports where men and women compete separately, has become a prime example of definitional uncertainty, as the understanding has widened, among biologists, that the binary notion of gender (male or female) is an inaccurate oversimplification.

In fact, gender is a spectrum,[4] for there are several ways in which its value may be expressed or assigned — based on anatomical features, hormonal profile, chromosomal structure, or self-identification —, which may contradict each other, giving rise to uncertainty.

[4] C. Ainsworth. Sex redefined. *Nature*, 518:288–291, February 2015. doi:10.1038/518288a. News Feature

These examples highlight how consideration of measurement uncertainty pervades not only areas of science and technology, but also many aspects of everyday life. Next we illustrate how measurement uncertainty can be propagated to the results of simple arithmetic calculations.

Sums, Products, and Ratios

In many cases, quantities of interest are expressed as sums, products, or ratios of quantities that have been measured previously. Such fairly simple measurement models serve to illustrate the basic procedures involved in uncertainty evaluations, including the propagation of uncertainties from input quantities to an output quantity, as in the following examples: (i) the plasma anion gap (expressed as a sum of four measured amount concentrations); (ii) the volume of a cylindrical storage tank (expressed as a product of two measured lengths); (iii) the resistance of an electric resistor (which is given by a ratio involving several measured resistances); and (iv) the atomic weight of lead (a sum of products).

Plasma Anion Gap

The plasma anion gap, Δc_{AG}, is used in clinical biochemistry to determine whether there is an imbalance of electrolytes in the blood, which may be a result of diabetes or of kidney disease, among other possibilities. It is defined as a linear combination of the amount concentration of two cations and two anions:

There are several different definitions of the anion gap. For example, it is common to omit potassium or to include corrections due to albumin.

$$\Delta c_{AG} = c(Na^+) + c(K^+) - c(Cl^-) - c(HCO_3^-).$$

Consider the values that were measured for a particular patient, shown in the table alongside. For this patient,

ION	c	$u(c)$
Na^+	137	1.48
K^+	4	0.04
Cl^-	106	0.72
HCO_3^-	10	0.84

Amount concentrations of ions (mmol/L) that were measured for a particular patient [White, 2008].

$$\Delta c_{AG} = (137 + 4 - 106 - 10)\,\text{mmol/L} = 25\,\text{mmol/L},$$

which generally would be regarded as being of clinical concern. However, the interpretation of any result of laboratory medicine requires consideration of the complete clinical profile of the patient [White et al., 2014] and requires also that measurement uncertainty be taken into account.

The uncertainty associated with the value of Δc_{AG} is determined by the reported uncertainties for the individual ion amount concentrations. These are the sizes of the margins of uncertainty discussed above, under *Measurement Uncertainty* (Page 14). White [2008] does not describe how they were evaluated, or which sources of uncertainty may have contributed to these values, but refers to them as standard deviations.

Following the *Guide to the expression of uncertainty in measurement* (GUM) [JCGM 100:2008], we refer to *standard uncertainty* to denote the standard deviation of the probability distribution that models the uncertainty surrounding the true value of the measurand.

We say that the standard uncertainty is associated with the measured value because different ways of measuring the same measurand may yield different uncertainties.

For this reason, $u(y)$ is the notation often used, where y is an estimate of the true value of the measurand, η.

However, the uncertainty is about η, not about y (whose value is known). The alternative, $u(\eta)$, would be logically appropriate but omits the dependence on the specific estimate. Maybe the notation $u_y(\eta)$ will gain traction some day.

This suggests that the underlying model for the measured amount concentrations involves random variables (Page 161) and probability distributions (Page 159), which provides a way forward to evaluate the standard uncertainty of the anion gap.

Indeed, if those four amount concentrations can be regarded as outcomes of independent random variables, then Δc_{AG} also is a random variable because it is a function of these random variables. Its variance (Page 165), denoted $u^2(\Delta c_{AG})$ below, can be computed exactly because the AG is a linear combination of the four amount concentrations, and the corresponding standard deviation, which will become its standard uncertainty, $u(\Delta c_{AG})$, is the square root of this variance:

$$\begin{aligned} u^2(\Delta c_{AG}) &= u^2(c(\mathrm{Na}^+)) + u^2(c(\mathrm{K}^+)) + \\ &\quad u^2(c(\mathrm{Cl}^-)) + u^2(c(\mathrm{HCO}_3^-)) \\ &= (1.48\,\mathrm{mmol/L})^2 + (0.04\,\mathrm{mmol/L})^2 + \\ &\quad (0.72\,\mathrm{mmol/L})^2 + (0.84\,\mathrm{mmol/L})^2 \\ &= (1.85\,\mathrm{mmol/L})^2. \end{aligned}$$

If an output quantity $Y = \alpha_1 X_1 + \cdots + \alpha_n X_n$ is a linear combination of uncorrelated input quantities for which estimates $x_1, \ldots, x_n$ and associated standard uncertainties $u(x_1), \ldots, u(x_n)$ are available, $\alpha_1, \ldots, \alpha_n$ are known constants, and $y = \alpha_1 x_1 + \cdots + \alpha_n x_n$, then $u^2(y) = \alpha_1^2 u^2(x_1) + \cdots + \alpha_n^2 u^2(x_n)$.

Even though Δc_{AG} involves sums and differences, the variances of the quantities being added or subtracted are all added (Page 167).

The precise meaning of $u(\Delta c_{AG}) = 1.85\,\mathrm{mmol/L}$ depends on the probability distribution of the random variable that is being used as a model for Δc_{AG}. If the four ion concentrations were modeled as Gaussian

It is a surprising fact that, for many probability distributions that a measurand y may have, the interval $y \pm 2u(y)$ will include the true value of y with approximately 95% probability [Freedman et al., 2007].

(Page 167) random variables, then so would be the Δc_{AG}, because a linear combination of independent (Page 163) Gaussian random variables is also Gaussian, and we would conclude that the true value of the Δc_{AG} lies between $(25 - 1.85)$ mmol/L and $(25 + 1.85)$ mmol/L with approximately 68 % probability.

Storage Tank

Consider the problem of evaluating and expressing the uncertainty that surrounds the internal volume V of a cylindrical storage tank, derived from measurement results for its radius R and for its height H. Since the volume is a nonlinear function of the radius and the height, $V = \pi R^2 H$, the form of calculation used for the anion gap does not apply to this case.

The volume of a cylindrical, oil storage tank is a nonlinear function of its height and diameter — PixelSquid (use licensed 2020).

The radius was measured by climbing a set of stairs to the tank's roof, whose shape and size are essentially identical to the shape and size of its base, measuring its diameter with a tape, and reporting the estimate of the radius as 8.40 m, give or take 0.03 m. This "give or take" is the margin of uncertainty.

One way to interpret this "give or take" involves modeling the measured value of the radius as $R + r$, where r denotes a measurement error, whose typical absolute value should be around 0.03 m, but that can be positive or negative. Still, without additional information or modeling assumption, this interpretation is not particularly meaningful and is insufficient to answer the question of how likely the true value of the radius is to lie between 8.37 m and 8.43 m, say. To answer this question one needs to give a probabilistic meaning to the concept of measurement uncertainty.

A modeling assumption that is commonly made is that r is like an outcome of a random variable whose expected value is 0 m and whose standard deviation is 0.03 m. This interpretation motivates calling 0.03 m *standard un-*

certainty. Alternatively, and equivalently, one could say that the measured value of the radius itself is like an outcome of a random variable with mean R and standard deviation 0.03 m.

What does the 0.03 m *actually subsume?* The standard uncertainty reflects contributions from all recognized sources of uncertainty that will have been evaluated individually and then combined to yield the reported value.

- Some of these contributions originate in the tape itself (how and when it was calibrated, or the effect that temperature has on its length);
- Other contributions derive from how the tape was laid out along a diameter of the roof (how stretched it may have been, how closely it passed to the actual center of the roof, and whether it touched and went over any rivets or ridges that may have made it deviate from a straight line parallel to the roof);
- Still other effects are attributable to how the tape was used by the person making the measurement (whether multiple measurements were made of the length of the diameter, and, if so, whether they were averaged or combined in some other way);
- And there will also be contributions from sources that are specific to the tank itself (how close to a perfect circle its roof may be, or how the temperature may affect the tank's volume and shape).

How likely is it that the true value of the radius indeed lies within 0.03 m *of the measured value, 8.40* m*?* To answer this question one needs a particular model for the uncertainty that the question alludes to. The kind of model that is used most often to address this question is a probabilistic model that characterizes in sufficient detail the random variable mentioned above. Such model is a *probability distribution*.

But which probability distribution? The answer depends on what is known about the sources of uncertainty listed above, and on how their contributions will have been combined into the reported margin of uncertainty.

A common choice (but by no means the best in all cases) is to use a Gaussian distribution (Page 167) as the model that lends meaning to the margin of uncertainty. In such case one can claim that the probability is about 68 % that the true value of the radius is within 0.03 m of its measured value.

The same questions need to be answered, and comparable modeling assumptions need to be made for the tank's height, H, which was measured using a plumb line dropped from the edge of the roof to the concrete platform that the tank is anchored to. The result turned out to be 32.50 m give or take 0.07 m.

Similarly to how we interpreted the result for the radius, here we regard the measured value of the height as $H + h$, where h denotes measurement error whose typical absolute value should be around 0.07 m.

The estimate of the volume,

$$V = \pi R^2 H = 7204\,\text{m}^3,$$

is obtained by substituting R and H by their measured values on the right-hand side of the measurement model for V.

Since the measured values of R and H are affected by errors, so will the resulting estimate of the volume. That is, $7204\,\text{m}^3 = V + v$, where v denotes the error that affects the measured value of the volume.

The error in the measured volume can be expressed as a function of the errors r and h that affect the measured values of the radius and of the height:

$$\begin{aligned} v &= \pi(R+r)^2(H+h) - V \\ &= \pi(R^2h + 2RHr + 2rhR + r^2H + r^2h) \\ &\approx \pi(R^2h + 2RHr). \end{aligned}$$

The approximation in the third line of this expression results from disregarding terms involving squares or products of the errors r and h, which is reasonable on the assumption that these errors are small.

If both r and h are regarded as outcomes of independent Gaussian random variables (Pages 163, 167 and 161), the third line of the expression above for v suggests that v also is approximately Gaussian.

Since products and squares of Gaussian random variables have distributions that are not Gaussian, the conclusion that v is approximately Gaussian is based only on the third line of the expression for v, after neglecting the terms in the second line of that expression that involve rh, r^2, and r^2h.

The approximation $v \approx \pi R^2 h + 2\pi RHr$, and the assumption that r and h are outcomes of independent random variables implies that the variance of v (that is, the square of its standard deviation, Page 165) is

$$\sigma_V^2 \approx (\pi R^2)^2\sigma_H^2 + (2\pi RH)^2\sigma_R^2,$$

where σ_R^2 and σ_H^2 denote the variances of r and h.

If we identify standard deviations with corresponding standard uncertainties, hence put $\sigma_R = u(R) = 0.03\,\text{m}$ and $\sigma_H = u(H) = 0.07\,\text{m}$, and otherwise substitute the symbols of the quantities above by their values, then we obtain

$$\begin{aligned} u^2(V) = \sigma_V^2 &\approx \left(\pi \times (8.40\,\text{m})^2\right)^2 \times (0.07\,\text{m})^2 + \\ &\quad \left(2\pi \times (8.40\,\text{m}) \times (32.5\,\text{m})\right)^2 \times (0.03\,\text{m})^2 \\ &\approx 2889\,\text{m}^6, \end{aligned}$$

hence $u(V) \approx 54\,\text{m}^3$. Next we will see that our intuitive error propagation exercise actually yielded the same approximate answer as a formula that Johann Carl Friedrich Gauss (1777–1855) was already using routinely, for the same purpose, in the first quarter of the 19th century, in his geodetic and astronomical work.

Gauss introduced the formula presented below as solution to the general problem in error propagation:[5]

[5] C. F. Gauss and G. W. Stewart. *Theory of the Combination of Observations Least Subject to Errors*. Classics in Applied Mathematics. SIAM (Society for Industrial and Applied Mathematics), Philadelphia, 1995. ISBN 978-0-89871-347-3. doi:10.1137/1.9781611971248

> Given a function U of the unknown quantities V, V', V'', etc., find the mean error M to be feared in estimating U when, instead of the true values of V, V', V'', etc. one uses independently observed values having mean errors m, m', m'', etc. — Gauss [1823, I.18]

Gauss's formula [Gauss, 1823] [Possolo and Iyer, 2017, VII.A.2], which is used in the *Guide to the expression of uncertainty in measurement* (GUM) [JCGM 100:2008], provides a practicable alternative that will produce a particularly simple approximation to the standard deviation of the output quantity because it is a product of powers of the input quantities: $V = \pi R^2 H^1$.

If an *output* quantity $y = f(x_1, \dots, x_n)$ is a function of n *input* quantities $x_1, \dots, x_n$ that have been measured with standard uncertainties $u(x_1), \dots, u(x_n)$, and the function f is differentiable, then $u^2(y) \approx (c_1 u(x_1))^2 + \dots + (c_n u(x_n))^2$, where c_j denotes the value that the first partial derivative of f with respect to x_j takes at $(x_1, \dots, x_n)$, for $j = 1, \dots, n$.

In this particular case, Gauss's formula says that the squared relative uncertainty of the volume has this particularly simple form:

$$\left(\frac{u(V)}{V}\right)^2 \approx \left(2\frac{u(R)}{R}\right)^2 + \left(1\frac{u(H)}{H}\right)^2.$$

Note that π does not figure in this formula because it has no uncertainty, and that the "2" and the "1" that appear as multipliers on the right-hand side are the exponents of R and H in the formula for the volume.

If the measurement model expresses the output quantity as $y = \kappa x_1^{\alpha_1} \dots x_n^{\alpha_n}$, where $\alpha_1, \dots, \alpha_n$ are (positive or negative) constants, and the standard uncertainties associated with the inputs are $u(x_1), \dots, u(x_n)$ such that $u(x_1)/x_1, \dots, u(x_n)/x_n$ are small ($< 10\,\%$), then $(u(y)/y)^2 \approx (\alpha_1 u(x_1)/x_1)^2 + \dots + (\alpha_n u(x_n)/x_n)^2$.

The approximation is likely to be quite accurate when the relative uncertainties, $u(R)/R$ and $u(H)/H$, are small (less than 10 %). Therefore,

$$u(V) \approx (7204\,\text{m}^3)\sqrt{4\left(\frac{0.03\,\text{m}}{8.40\,\text{m}}\right)^2 + \left(\frac{0.07\,\text{m}}{32.50\,\text{m}}\right)^2} = 54\,\text{m}^3,$$

which is precisely the same that we obtained in our intuitive error propagation exercise described above.

A MONTE CARLO METHOD [Possolo and Iyer, 2017, VII.A.3] for uncertainty propagation introduced by Morgan and Henrion [1992] and described in JCGM 101:2008, provides yet another eminently practicable alternative, whose validity does not depend on the relative standard uncertainties being small. The idea and execution both are very simple:

(1) Make a large number ($K \approx 10^6$) of drawings from the probability distributions of R and H, using their measured values as the means of these distributions, and their reported standard uncertainties as the standard deviations.

(2) For each pair of these draws, R_k and H_k, calculate the volume of the cylinder $V_k = \pi R_k^2 H_k$, for $k = 1, \ldots, K$.

(3) Calculate the average of these volume values, $V_1, \ldots, V_K$, and use it as an estimate of the mean value of V, and their standard deviation as an estimate of $u(V)$.

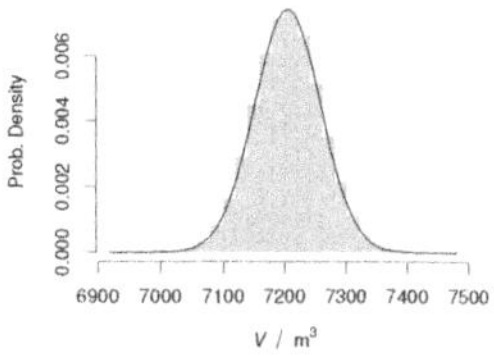

Histogram of 10^6 replicates of the value of V simulated using the Monte Carlo method, and probability density (smooth curve) of the Gaussian distribution with the same mean and standard deviation as those replicates.

Using samples of size $K = 10^6$, we reached the conclusion that $V = 7204\,\text{m}^3$, give or take $54\,\text{m}^3$, and the histogram of these one million replicates shows that V has a probability density that is virtually indistinguishable from the density of a Gaussian distribution with mean $7204\,\text{m}^3$ and standard deviation $54\,\text{m}^3$. Note, however, that in general the probability distribution of the output quantity need not be close to Gaussian, even when the distributions of the input quantities are Gaussian.

Wheatstone Bridge

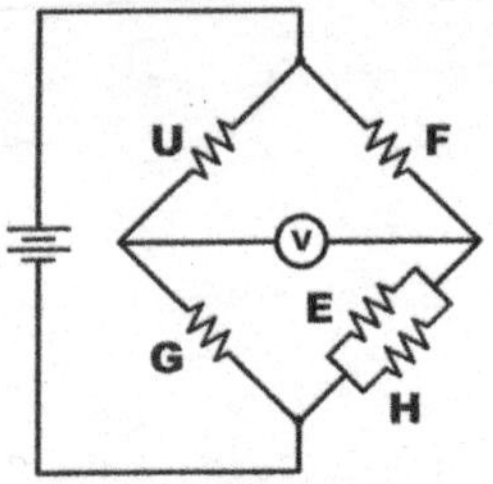

Wheatstone bridge comprising the resistor U whose resistance, R_U, one intends to measure, a resistor F with fixed resistance, and three resistors (G, E, and H) with adjustable resistances.

The Wheatstone bridge is an electrical circuit used to obtain accurate measurements of resistance by balancing both sides of a bridge circuit, one of which includes the component with unknown resistance (resistor U). In its simplest version, the Wheatstone bridge comprises a DC power supply, a voltmeter, and four resistors, one of which has adjustable resistance. The bridge illustrated here comprises three adjustable resistors, two of which are arranged in parallel so as to achieve finer control over their joint resistance, which is half the harmonic mean of their individual resistances, R_E and R_H:

$$R_{EH} = \frac{1}{R_E^{-1} + R_H^{-1}}.$$

The choice of instrumentation described above pays homage to a bygone era of analog electrical devices. The General Radio Company designed and manufactured test equipment for resistance, inductance, and capacitance, from 1915 until 2001, in West Concord, MA. The Electronic Instrument Company (EICO) was established in Brooklyn NY, in 1945, and remained in business for 54 years. Besides test equipment, EICO also produced Geiger counters, as well as amateur radio and high-fidelity audio equipment.

Resistor G is a General Radio decade resistor that can take values of resistance up to 1 MΩ in increments of 0.1 Ω, with relative standard uncertainty 0.05 %. Resistor E is an EICO decade resistor that can take values up to 100 kΩ in increments of 1 Ω, with relative standard uncertainty 0.5 %, and resistor H is a Heathkit RS-1 Resistance Substitution Box that allows the user to select one of several values of resistance.

We assume that the measurement experiment was carried out quickly enough, and at sufficiently low voltage (4 V), so that changes in resistance caused by heating of the resistors are negligible. We also assume that the error is negligible in achieving 0 V when balancing the bridge by adjusting the resistances of G, E, and H, thus reaching the point where R_U/R_G equals R_F/R_{EH}. Hence, we have the following measurement equation for R_U:

$$R_U = \frac{R_G R_F}{R_{EH}} = R_G R_F \left(R_E^{-1} + R_H^{-1}\right)$$

	R	$u(R)$
E	951 Ω	5 Ω
F	997 Ω	5 Ω
G	909 Ω	0.5 Ω
H	225.2 kΩ	2.3 kΩ

Observed resistance values that result in zero volt potential difference across the Wheatstone bridge.

The observed resistance values with the associated standard uncertainties are listed in the table alongside.

Since R_U is not a simple product of powers of R_E, R_F, R_G, and R_H, the approximation used above, for the uncertainty of the volume of the storage tank, cannot be used here. For this we use Gauss's formula in its general form, which relates the uncertainties associated with uncorrelated input quantities R_E, R_F, R_G, and R_H, with the output quantity R_U:

$$u^2(R_U) \approx \left(\frac{\partial R_U}{\partial R_E}\right)^2 u^2(R_E) + \left(\frac{\partial R_U}{\partial R_F}\right)^2 u^2(R_F) + \left(\frac{\partial R_U}{\partial R_G}\right)^2 u^2(R_G) + \left(\frac{\partial R_U}{\partial R_H}\right)^2 u^2(R_H).$$

DERIV.	VALUE
$\partial R_U/\partial R_E$	$-R_G R_F / R_E^2$
$\partial R_U/\partial R_F$	$R_G(R_E^{-1} + R_H^{-1})$
$\partial R_U/\partial R_G$	$R_F(R_E^{-1} + R_H^{-1})$
$\partial R_U/\partial R_H$	$-R_G R_F / R_H^2$

Partial derivatives of the output quantity, R_U, with respect to all four input quantities. Note that the expression in the second line is R_U/R_F, and the one the third line is R_U/R_G. These and other derivatives can be readily obtained using a variety of online tools such as www.wolframalpha.com.

The partial derivatives of the measurement model are given in the table alongside. By substituting them into the expression above, we obtain

$$u^2(R_U) = \frac{R_G^2 R_F^2}{R_E^4} u^2(R_E) + \frac{R_U^2}{R_F^2} u^2(R_F) + \frac{R_U^2}{R_G^2} u^2(R_G) + \frac{R_G^2 R_F^2}{R_H^4} u^2(R_H).$$

Finally, the estimate of the measurand is

$$R_U = 909\,\Omega \times 997\,\Omega \times \left(\frac{1}{951\,\Omega} + \frac{1}{225.2\,\mathrm{k}\Omega}\right) = 957\,\Omega,$$

with associated standard uncertainty $u(R_U) \approx 7\,\Omega$.

The *NIST Uncertainty Machine* [Lafarge and Possolo, 2015] can produce these results in a single stroke. Modeling all the resistances as Gaussian random variables with means equal to the observed values and standard deviations equal to the standard uncertainties, we obtain not only $R_U = 957\,\Omega$ and $u(R_U) = 7\,\Omega$, but also a probability distribution for R_U and, in turn, a 95 % coverage interval for the true value of R_U. We also learn that the squared uncertainties of the resistors, $u^2(R)$, contribute to the $u^2(R_U)$ in these proportions: F: 48 %; G: 0.6 %; E: 52 %; and H: 0.004 %.

Resistance is a positive quantity while the Gaussian uncertainty model entertains the possibility of negative values. For this reason, the lognormal (Page 172) model is sometimes chosen. The Gaussian and lognormal models are just about identical when the relative uncertainties are small ($< 5\,\%$).

[6] H. Zangl, M. Zine-Zine, and K. Hoermaier. Utilization of software tools for uncertainty calculation in measurement science education. *Journal of Physics: Conference Series*, 588:012054, 2015. doi:10.1088/1742-6596/588/1/012054

The measurement model considered above does not recognize the uncertainty associated with balancing the Wheatstone bridge. A more elaborate model that accounts for this is as follows:[6]

$$R_U = \frac{U_0 R_G (R_F + R_{EH})}{U_0 R_{EH} + U(R_F + R_{EH})} - R_G.$$

Here, U_0 is the potential difference across the terminals of the DC power supply, $U_0 = 4\,\text{V}$, and U is the potential across the balanced bridge ($U \approx 0\,\text{V}$). Uncertainty analysis of this more complete measurement model using the *NIST Uncertainty Machine* reveals that balancing the Wheatstone bridge becomes the dominant source of uncertainty of R_U if the uncertainty associated with U is larger than 5 mV.

Atomic Weight of Lead

Stefanie Horowitz (1887–1942) demonstrated experimentally that the same element can have different atomic weights depending on its source [Hönigschmid et al., 1915]. Her meticulous gravimetric analysis showed that lead from the pitchblende in which the Curies had discovered polonium and radium had a much lower atomic weight (206.6) than common lead (207.2).

The atomic weight of lead in a sample of a material is the average of the relative atomic masses of all the atoms of lead in the sample. This average can be different for different materials because there are four different atoms of lead, with different relative atomic masses, and their proportions vary between materials. These atoms are the isotopes ^{204}Pb, ^{206}Pb, ^{207}Pb, and ^{208}Pb.

Two materials with extreme atomic weights of lead are specimens of the mineral monazite from metamorphic rocks in Scotland that are more than 2.5 billion years old: the atomic weight of lead in one of them is 206.1462 ± 0.0028, and in the other it is 207.9351 ± 0.0005, where the quoted uncertainties are expanded uncertainties for 95 % coverage [Zhu et al., 2021].

The GUM defines *expanded uncertainty* as a multiple of the standard uncertainty, and denotes it with the uppercase letter U, as in $U(y) = ku(y)$, and calls the multiplier, k, the *coverage factor*. This coverage factor is selected so that $y \pm U(y)$ includes the true value of y with a specified probability, for example, 95 %. It is customary to write this probability as a subscript for the uppercase U, as in $y \pm U_{95\,\%}(y)$.

When Monte Carlo methods are used for uncertainty evaluation, expanded uncertainties are typically obtained directly rather than via "expansion" of the standard uncertainty.

Tong et al. [2019] obtained the following measurement results for three isotopic ratios in the reference material HIPB-1, a lead wire whose isotopic composition was certified by the National Research Council Canada [Meija et al., 2020]:

	$^{204}\mathrm{Pb}/^{207}\mathrm{Pb}$	$^{206}\mathrm{Pb}/^{207}\mathrm{Pb}$	$^{208}\mathrm{Pb}/^{207}\mathrm{Pb}$
R	0.063 00	1.3314	2.5193
$U_{95\,\%}(R)$	0.000 06	0.0004	0.0008

The row labeled R has the values of the ratios, and the line labeled $U_{95\,\%}(R)$ has the associated expanded uncertainties corresponding to the coverage factor $k = 2$.

These isotope ratios, R, were determined using multi-collector inductively coupled plasma mass spectrometry, and were then used to estimate the atomic weight of lead in this material and to evaluate the associated uncertainty, which involves the following steps:

(1) Compute the amount fractions $x(^{204}\mathrm{Pb})$, $x(^{206}\mathrm{Pb})$, $x(^{207}\mathrm{Pb})$, and $x(^{208}\mathrm{Pb})$ that are consistent with these ratios and satisfy the constraints of being non-negative and adding to 1. For example,

$$x(^{204}\mathrm{Pb}) = \frac{R_{204/207}}{R_{204/207} + R_{206/207} + 1 + R_{208/207}}.$$

(2) Evaluate the standard uncertainties and correlations associated with these amount fractions.

(3) Compute the atomic weight of lead in this material, $A_r(\mathrm{Pb})$, as a weighted average of the relative atomic masses of these four isotopes, with those amount fractions as weights:

$$\begin{aligned} A_r(\mathrm{Pb}) = & A_r(^{204}\mathrm{Pb})x(^{204}\mathrm{Pb}) + A_r(^{206}\mathrm{Pb})x(^{206}\mathrm{Pb}) + \\ & A_r(^{207}\mathrm{Pb})x(^{207}\mathrm{Pb}) + A_r(^{208}\mathrm{Pb})x(^{208}\mathrm{Pb}). \end{aligned}$$

(4) Propagate the uncertainties and correlations for the amount fractions, and the uncertainties associated with the relative atomic masses of the isotopes, to obtain $u(A_r(\mathrm{Pb}))$ that is associated with $A_r(\mathrm{Pb})$ in HIPB-1.

Before taking these steps, we point out that the data reductions we will describe produce some results that differ slightly from those listed in the certificate of HIPB-1, even if the atomic weight of lead reproduces the value and uncertainty in the certificate.[7] The main reason for the differences is that we use slightly different data reduction methods. Users of HIPB-1 should rely on the values listed in the certificate.

[7] J. Meija, B. Methven, S. Tong, O. Mihai, K. Swider, P. Grinberg, Z. Mester, and L. Yang. HIPB-1: High Purity Lead Certified Reference Material for Lead Mass Fraction, Atomic Weight, Isotopic Composition and Elemental Impurities. National Research Council Canada, Ottawa, 2020

Step (1) involves nonlinear constrained optimization carried out using R function `solnp` defined in package `Rsolnp`, which implements an augmented Lagrange multiplier method,[8] using the following R code:

[8] Y. Ye. *Interior Point Algorithms: Theory and Analysis*. John Wiley & Sons, New York, NY, 1997. ISBN 978-0471174202; and A. Ghalanos and S. Theussl. *Rsolnp: General Non-linear Optimization Using Augmented Lagrange Multiplier Method*, 2015. R package version 1.16

```
require(Rsolnp)
require(mvtnorm)

negLogLik = function (x, r, sigma) {
    Pb204 = x[1]
    Pb206 = x[2]
    Pb207 = x[3]
    Pb208 = x[4]
    rho = c(Pb204/Pb207, Pb206/Pb207, Pb208/Pb207)
    return(-1*dmvnorm(r, mean=rho, sigma=sigma, log=TRUE))
    }

r  = c(0.06300, 1.3314, 2.5193)
ur = c(0.00006, 0.0004, 0.0008)/2
sigma.r = diag(ur^2)

mle = solnp(pars=c(0.012822, 0.27096, 0.203511, 0.51271),
         fun=negLogLik,
         eqfun=function(x,r,sigma){ sum(x) },
         eqB=1, LB=rep(0,4), UB=rep(1,4),
         r=r, sigma=sigma.r)
xHAT = mle$pars
```

We employ a Monte Carlo method in step (2), under the simplifying assumption that the isotope ratios are like outcomes of independent Gaussian random variables. Subsequently, we recognize the correlations between the amount fractions that are attributable to the so-called *closure effect*,[9] that is, to the fact that the amount fractions have a constant sum. The following block of R code implements this second step.

[9] F. Chayes. *Ratio Correlation: A Manual for Students of Petrology and Geochemistry*. University of Chicago Press, Chicago, Illinois, 1971

```
K = 50000
rB = rmvnorm(K, mean=r, sigma=sigma.r)
xB = array(rep(NA,4*K), dim=c(K,4))
for (k in 1:K) {
  mleB = solnp(pars=xHAT, fun=negLogLik,
               eqfun=function(x,r,sigma){ sum(x) },
               eqB=1, LB=rep(0,4), UB=rep(1,4),
               control=list(trace=0),
               r=rB[k,], sigma=sigma.r)
  xB[k,] = mleB$pars
}
```

The standard uncertainties for the isotopic abundances of lead in HIPB-1 are

	^{204}Pb	^{206}Pb	^{207}Pb	^{208}Pb
$u(x)/(\text{mol/mol})$	0.000 006	0.000 037	0.000 019	0.000 045

	^{204}Pb	^{206}Pb	^{207}Pb
^{206}Pb	−0.02		
^{207}Pb	+0.13	+0.18	
^{208}Pb	−0.17	−0.90	−0.58

and the corresponding correlation matrix is alongside.

Finally, in step (3), we resort to the Monte Carlo method for a second time to propagate not only the standard uncertainties and correlations of the amount fractions, but also the uncertainties associated with the relative atomic masses of the isotopes, which are as follows:[10]

[10] M. Wang, W. J. Huang, F. G. Kondev, G. Audi, and S. Naimi. The AME 2020 atomic mass evaluation (II). Tables, graphs, and references. *Chinese Physics C*, 45(3):030003, 2021. doi:10.1088/1674-1137/abddaf

Isotope	A_r	$u(A_r)$
^{204}Pb	203.973 0435	0.000 0012
^{206}Pb	205.974 4652	0.000 0012
^{207}Pb	206.975 8968	0.000 0012
^{208}Pb	207.976 6520	0.000 0012

```
Ar   = c(Pb204=203.9730435, Pb206=205.9744652,
         Pb207=206.9758968, Pb208=207.9766520)
Ar.u = c(Pb204=0.0000012,   Pb206=0.0000012,
         Pb207=0.0000012,   Pb208=0.0000012)
K = nrow(xB)
ArB = matrix(rnorm(4*K, mean=Ar, sd=Ar.u), nrow=K, byrow=TRUE)
ArPb = apply(ArB*xB, 1, sum)
c(mean(ArPb), diff(quantile(ArPb, probs=c(0.025,0.975)))/2)
```

Since the 1960s, atomic masses of nuclides are expressed relative to 1/12th of the mass of the carbon-12 atom. This atomic mass unit is called "dalton" (symbol Da) in honor of the English chemist John Dalton (1766–1844), who introduced the atomic theory in chemistry.

The resulting estimate of $A_r(\text{Pb})$ in the certified reference material HIPB-1 is 207.1791, and the expanded uncertainty for 95 % coverage associated with this estimate is $U_{95\,\%}(A_r(\text{Pb})) = 0.0002$.

Monitoring

The evaluation of measurement uncertainty is key to determining whether a process evolving over time remains stable, or undergoes changes. In 1924, Walter Shewhart, then working for the Western Electric Company, developed a graphical display to monitor the quality of telephone components being manufactured at the company's Hawthorne Works (Cicero, Illinois).[11]

[11] W. A. Shewhart. *Economic Control of Quality of Manufactured Product.* D. Van Nostrand Company, Princeton, NJ, 1931

Such graphical displays, which became known as *control charts*, serve to monitor production processes in real-time, and to suggest when process adjustments may be required to ensure that values of selected properties of product quality remain within specified control limits. Shewhart attributed variations in product quality either to *chance causes* or to *assignable causes*.[12] The former reflect the natural variability of a production process, while the latter are the consequence of identifiable disturbances of the process.

[12] R. E. Barlow and T. Z. Irony. Foundations of statistical quality control. In M. Ghosh and P. K. Pathak, editors, *Current issues in statistical inference: Essays in honor of D. Basu*, volume 17 of *IMS Lecture Notes – Monograph Series*, pages 99–112. Institute of Mathematical Statistics, 1992. ISBN 0-940600-24-2. doi:10.1214/lnms/1215458841

The first example in this section, concerning the determination of silver impurities in a copper rod, demonstrates how control charts can be used to monitor the stability of repeated measurements, made over time, of an invariant measurand, and to detect changes in measured values that indicate the need for recalibration or some other adjustment of the measuring instrument. In this case, miscalibration is an assignable cause that induces excessive variability of the measured values.

Once assignable causes are identified and removed, the process is in statistical control. "A process is said to have reached a state of statistical control when changes in measures of variability and location from one sampling period to the next are no greater than statistical theory would predict."[13]

[13] L. S. Nelson. Control charts. In S. Kotz, N. Balakrishnan, C. B. Read, B. Vidakovic, and N. L. Johnson, editors, *Encyclopedia of Statistical Sciences*. John Wiley & Sons, Hoboken, NJ, second edition, 2005. ISBN 978-0-471-15044-2

The second example involves a series of measurements of temperature made to characterize the thermal stability of a water bath used for the calibration of thermometers.

Such stability is ascertained qualitatively by establishing that a stationary time series model is adequate for the series of temperature values, and then it is quantified by the standard uncertainty associated with the estimate of the mean of a first-order auto-regression with Gaussian innovations.

The third example concerns the temporal evolution of the amount fraction of carbon dioxide in the Earth's atmosphere during the most recent two thousand years: not only has this value changed dramatically, especially since the beginning of the Industrial Revolution (1760s), but the very model that describes such changes, a Gaussian process regression, has had to change, too, to accommodate structural changes to patterns of anthropogenic emissions that took place toward the end of the 16th century and, far more dramatically and enduringly, since the Industrial Revolution.

Silver Impurities

The National Research Council Canada has been measuring the mass fraction of silver in a copper rod (certified reference material BCR-075B) since January 2005, using glow discharge mass spectrometry (GDMS),[14] to monitor the stability of the measuring instrument. The measurements made until the end of 2021 have been irregularly spaced in time, with waiting time between consecutive measurements around 21 days, give or take 9 days.

[14] V. Hoffmann, M. Kasik, P. K. Robinson, and C. Venzago. Glow discharge mass spectrometry. *Analytical and Bioanalytical Chemistry*, 381:173–188, 2005. doi:10.1007/s00216-004-2933-2

The Shapiro-Wilk[15] test of Gaussian shape suggests that the $n = 256$ determinations made in the course of the intervening two decades do not appear to be a sample from a Gaussian distribution, because the probability is only 0.001 of observing a sample at least as deviant from the Gaussian "standard" as this one.

[15] S. S. Shapiro and M. B. Wilk. An analysis of variance test for normality (complete samples). *Biometrika*, 52(3,4):591–611, 1965. doi:10.2307/2333709

The dubious logic behind this conclusion is that rare events should not occur, and if they do occur, then one should question the validity of the assumption that renders the event rare. In other words, and in this case, if the sample we have observed is most unusual, then it is most unlikely that the distribution the sample comes from is Gaussian.

The above probability of 0.001 is called the p-value of the test. Since it is quite small on the assumption that the sample comes from a Gaussian distribution, this possibility is dismissed while allowing that such a decision may be erroneous with that same probability. Harold Jeffreys characterized this behavior as follows [Jeffreys, 1961, Page 385]:

> What the use of P implies, therefore, is that a hypothesis that may be true may be rejected because it has not predicted observable results that have not occurred.

A robust estimate of the center of the probability distribution of the mass fraction of silver in this material is $\widetilde{w} = 12.40\,\mu\text{g/g}$, with the associated standard uncertainty $u(\widetilde{w}) = 0.05\,\mu\text{g/g}$. This estimate, which was obtained by applying R function `huberM` defined in package `robustbase`[16] to the 256 determinations of the mass fraction of silver in the copper rod, determines the position of the horizontal thin line in the following charts.

The first chart is a simplified, robust version of the Shewhart control chart for individual measurements[17].

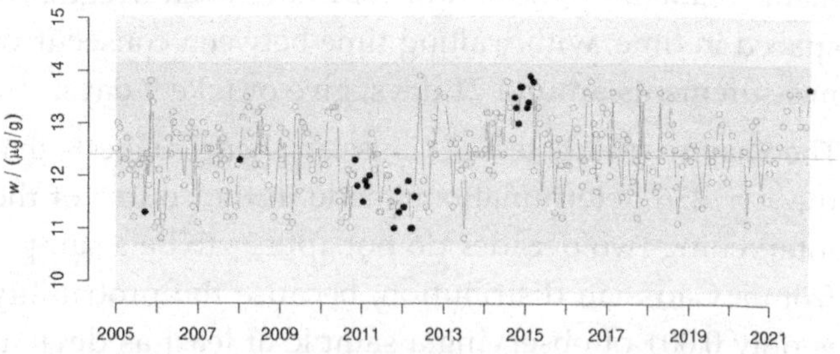

The circles (outline or solid) represent the measured values, the height of the light gray band represents $\widetilde{w} \pm 3\widetilde{\sigma}$, and the height of the dark gray band represents $\widetilde{w} \pm 2\widetilde{\sigma}$, where $\widetilde{\sigma} = 0.88\,\mu\text{g/g}$ is a robust estimate of the

[16] M. Maechler, P. Rousseeuw, C. Croux, V. Todorov, A. Ruckstuhl, M. Salibian-Barrera, T. Verbeke, M. Koller, E. L. T. Conceição, and M. A. di Palma. *robustbase: Basic Robust Statistics*, 2021. URL `http://CRAN.R-project.org/package=robustbase`. R package version 0.93-9

[17] NIST/SEMATECH. *NIST/SEMATECH e-Handbook of Statistical Methods*. National Institute of Standards and Technology, U.S. Department of Commerce, Gaithersburg, Maryland, 2012. doi:10.18434/M32189. URL `https://www.itl.nist.gov/div898/handbook/`

standard deviation of the observations, computed using R function `Qn`[18] defined in package `robustbase`.

[18] P. J. Rousseeuw and C. Croux. Alternatives to the median absolute deviation. *Journal of the American Statistical Association*, 88:1273–1283, December 1993

The solid circles indicate measured values that trigger at least one of the Western Electric decision rules indicating anomalous patterns in control charts:[19] in this case, instances of at least eight consecutive points that are all above, or all below the center line.

[19] Western Electric. *Statistical Quality Control Handbook*. Western Electric Corporation, Indianapolis, IN, 2nd edition, 1958

The Exponentially Weighted Moving Average (EWMA) chart depicts (solid dark gray circles joined by lines) the value of a weighted average, of the value at each particular epoch and of the previous values, with weights that decay exponentially fast toward zero according to how far back in time the previous values are.

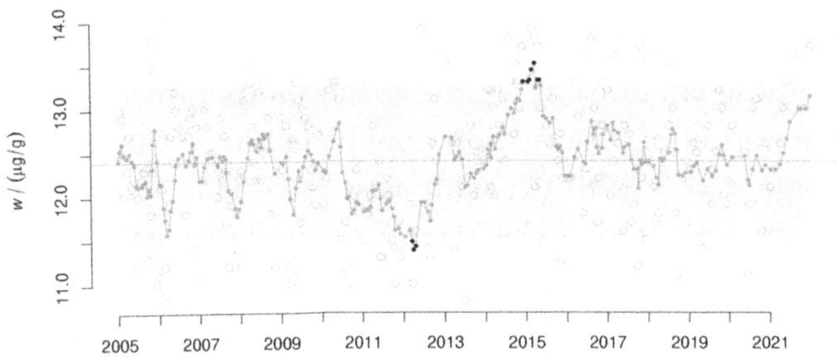

More precisely, the EWMA values are

$$w^*(t_i) = \lambda w(t_i) + (1 - \lambda) w^*(t_{i-1}),$$

with $t_{i-1} < t_i$, for $i = 1, \ldots, n$, and $w^*(t_0)$ set equal to the aforementioned $\widetilde{w} = 12.40\,\mu\text{g/g}$. The "memory" parameter λ has been set equal to 0.2 as recommended by Hunter [1986], and this value is also the default choice in the implementation in R function `ewma`, which is defined in package `qcc`.[20]

[20] L. Scrucca. qcc: an R package for quality control charting and statistical process control. *R News*, 4:11–17, 2004. URL `https://cran.r-project.org/doc/Rnews/`

The EWMA chart highlights periods with significant changes in the mean of the process: one around the beginning of 2012, when the GDMS instrument was re-calibrated, and another in 2015.

Thermal Bath

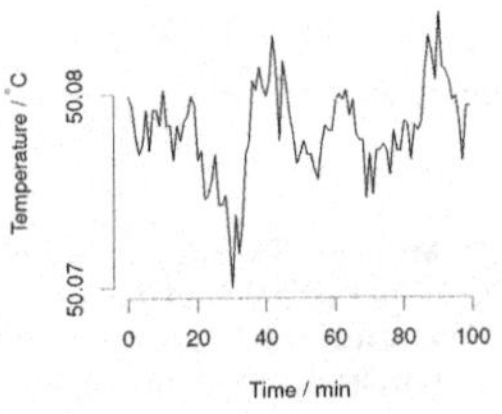

The readings of the temperature of a thermal bath that are depicted alongside and listed below, were taken every minute in the course of 100 min with a thermocouple immersed in the bath to ascertain that the temperature of the bath remained stable for the duration, and to measure its mean temperature.

These readings (which are listed in the R code below) were made under conditions of repeatability, and may reasonably be regarded as a sample from a Gaussian probability distribution (Page 167): the Anderson-Darling [Anderson and Darling, 1952] test of Gaussian shape yields a p-value of 0.13.

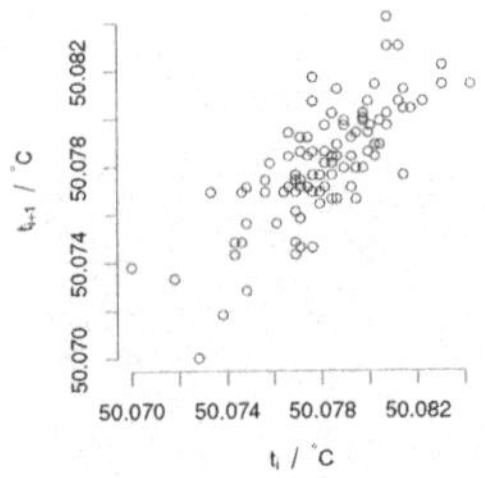

The coordinates of each point are consecutive readings of temperature of the thermal bath.

In these circumstances, the mean temperature during the period of observation may be estimated by the average, $\bar{t} = 50.0781\,°\mathrm{C}$, with associated standard uncertainty evaluated by the Type A method proposed in the GUM 4.2.3 [JCGM 100:2008], as

$$u(\bar{t}) = 0.0024\,°\mathrm{C}/\sqrt{100} = 0.000\,24\,°\mathrm{C}$$

because the sample standard deviation of these 100 readings is 0.0024 °C.

However, this evaluation of $u(\bar{t})$ assumes that the readings are uncorrelated, an assumption that the plot alongside reveals to be unrealistic: there is a strong tendency for consecutive values of temperature to be positively correlated.

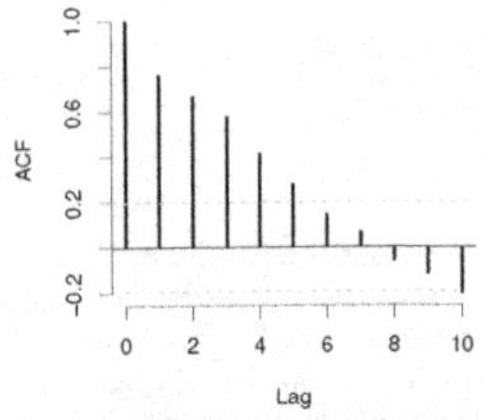

Autocorrelation function (ACF) for series of temperature readings: the value of the ACF at lag ℓ, represented by a vertical line segment, is the correlation coefficient of the pairs $\{t_i, t_{i+\ell}\}$. Those line segments that extend beyond the horizontal, dashed lines, differ significantly from 0.

The autocorrelation function (depicted alongside) shows that such autocorrelations are pervasive in this series of readings of temperature.

Equation (16) in the GUM provides a way to take correlations into account, which involves estimating each one of them. Doing so in this case would give us a rather mixed bag of estimates, in terms of their reliability, because some autocorrelations would be estimated based on many pairs of readings (those that correspond to small lags), while others would be based on only a few (those others that correspond to large lags).

Instead, we take a different approach, which involves modeling the time series of readings, and then deriving from this model an evaluation of $u(\bar{t})$ without invoking the approximation in the GUM Equation (16). The model we shall entertain will be selected from the class of autoregressive-moving-average (ARMA) models [Box et al., 2008].

Here we use R function `auto.arima`, defined in package `forecast`,[21] and the Bayesian Information Criterion (BIC, Page 100), to select the best model in the subset of that class that includes ARMA models with no more than 10 autoregressive or moving average parameters in total.

[21] R.J. Hyndman and Y. Khandakar. Automatic time series forecasting: the forecast package for R. *Journal of Statistical Software*, 27:1–22, July 2008

```
t = 50 + c(799, 794, 779, 769, 774, 792, 771, 792, 792, 784, 802,
784, 784, 766, 784, 776, 786, 789, 799, 794, 766, 771, 746, 748,
756, 769, 743, 743, 748, 728, 700, 738, 718, 733, 769, 776, 807,
802, 814, 804, 799, 807, 830, 814, 776, 817, 804, 789, 779, 764,
769, 776, 769, 769, 761, 756, 774, 784, 781, 781, 797, 800, 797,
802, 789, 797, 779, 776, 776, 746, 769, 748, 771, 771, 774, 771,
758, 781, 771, 771, 786, 784, 766, 784, 781, 786, 812, 830, 822,
807, 842, 814, 812, 807, 797, 799, 786, 766, 794, 794)/10000

library(forecast)
z = auto.arima(t, stepwise=FALSE, max.order=10, ic="bic",
       stationary=FALSE, seasonal=FALSE, approximation=FALSE)
summary(z)
```

The selected model is a first order, stationary autoregression

$$t_{i+1} = \tau + \varphi(t_i - \tau) + \varepsilon_{i+1}$$

where τ denotes the true mean temperature of the bath, φ is the autoregressive parameter, and the $\{\varepsilon_i\}$ are non-observable outcomes of mutually independent Gaussian random variables with mean 0 and standard deviation σ. The maximum likelihood estimates (Page 191) of the model parameters are $\widehat{\tau} = 50.0782\,°\mathrm{C}$, $\widehat{\varphi} = 0.7631$, and $\widehat{\sigma} = 0.001\,551\,°\mathrm{C}$.

The model says that each reading t_{i+1} comprises the

additive superposition of a fraction of the deviation of the previous reading from the common mean τ, $\varphi(t_i - \tau)$, and a contribution from white noise, in the form of the Gaussian "innovation" ε_{i+1}.

Since $\widehat{\varphi} > 0$, the model describes the tendency for consecutive readings to be positively correlated. Because $-1 < \widehat{\varphi} < +1$, the sequence of temperature readings is *stationary*, which implies that the bath is in thermal equilibrium. In particular, this means that all the $\{t_i\}$ have the same expected value τ and the same standard deviation $\sigma/\sqrt{1-\varphi^2}$, and that the correlation between t_i and $t_{i+\ell}$ depends only on the value of ℓ (which may be positive or negative).

The R code above produces an evaluation of $u(\widehat{\tau}) = 0.000\,64\,°\mathrm{C}$, almost 3 times larger than the naive evaluation of $u(\bar{t})$ given above, which neglected the correlations between the readings.

Atmospheric Carbon Dioxide

The World Data Service for Paleoclimatology (Boulder, Colorado), and the NOAA Paleoclimatology Program at the National Centers for Environmental Information make available dataset with values of the amount fraction of CO_2 in air bubbles preserved in an ice core drilled at the Law Dome, Antarctica, and corresponding estimates of the age of the CO_2.[22]

[22] M. Rubino, D. M. Etheridge, D. P. Thornton, R. Howden, C. E. Allison, R. J. Francey, R. L. Langenfelds, L. P. Steele, C. M. Trudinger, D. A. Spencer, M. A. J. Curran, T. D. van Ommen, and A. M. Smith. Revised records of atmospheric trace gases CO_2, CH_4, N_2O and $\delta^{13}C$-CO_2 over the last 2000 years from Law Dome, Antarctica. *Earth System Science Data*, 11(2):473–492, 2019. doi:10.5194/essd-11-473-2019

A *treed Gaussian process regression* was fitted to the data from the Law Dome, using the following R code.

The antarctic ice core data are available at ncei.noaa.gov/access/paleo-search/study/25830

```
require(tgp)
URL = paste0("https://www.ncei.noaa.gov/pub/data/",
          "paleo/icecore/antarctica/law/law2018co2.txt")
co2 = read.table(url(URL), header=TRUE, sep = "\t")

gp = btgp(X=co2$age_CO2, Z=co2$CO2ppm,
          XX=seq(from=min(co2$age_CO2), to=max(co2$age_CO2), by=10),
          bprior="b0", tree=c(0.5, 2), BTE=c(5000, 20000, 10) )
plot(gp); gp$trees[[3]]$val[1:2]; tgp.trees(gp)
```

The function `btgp` defined in R package `tgp` [Gramacy, 2007] implements a Bayesian procedure that was used to fit the model to the data from the Law Dome.[23] [24]

[23] R. B. Gramacy and H. K. H. Lee. Bayesian treed Gaussian process models with an application to computer modeling. *Journal of the American Statistical Association*, 103:1119–1130, 2008

[24] H. Chipman, E. I. George, R. B. Gramacy, and R. McCulloch. Bayesian treed response surface models. *Wiley Interdisciplinary Reviews: Data Mining and Knowledge Discovery*, 3(4):298–305, 2013. doi:10.1002/widm.1094

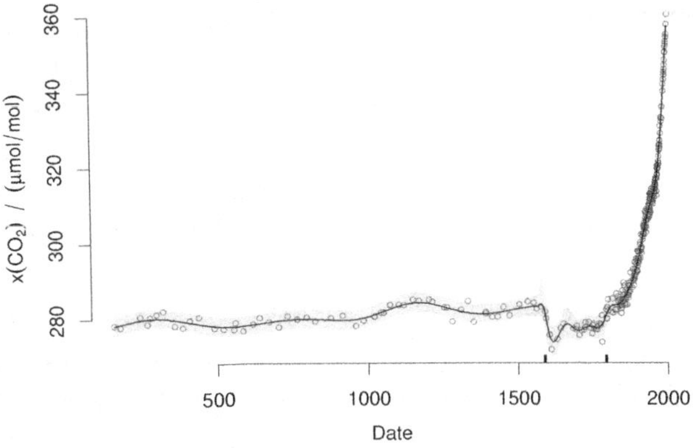

The measured values and the fitted model are depicted in the figure above. The figure also shows a 95 % coverage band for the fitted curve, which depicts the means of three Gaussian processes (Page 93) joined end-to-end.

The model identified two dates, 1588 and 1794, when the structure of the data changed, which required switching from one Gaussian process to another. These dates, which are marked in the figure by thick, short tick marks pointing up from the horizontal axis, partition the data into three periods. During the first period, which lasted for more than one thousand years, the amount fraction of CO_2 remained stable, hovering around 280 μmol/mol.

Theaters remained closed 1592–1593, but when they reopened the following year, William Shakespeare had both a new tragedy, *Titus Andronicus*, and a new comedy, *The Taming of the Shrew*, ready for performance.

The staggering increase of the amount fraction of CO_2 during the third period, from 1790s onward, is the consequence of the ever increasing anthropogenic emissions of CO_2 that started with the Industrial Revolution.

The dip in the amount fraction of CO_2 starting in the 1590s may well be a consequence of the plague epidemic in England during the late 1580s and early 1590s which claimed about 13 % of the population of London, and that was also followed by forest regrowth.[25]

[25] T. van Hoof, F. P. M. Bunnik, J. G. M. Waucomont, W. M. Kürschner, and H. Visscher. Forest re-growth on medieval farmland after the black death pandemic — implications for atmospheric CO_2 levels. *Palaeogeography, Palaeoclimatology, and Palaeoecology*, 237:396–409, 2006. doi:10.1016/j.palaeo.2005.12.013

Counting

Counting may be the simplest form of measuring. The value we assign to a count is based on comparisons with two kinds of standards. One standard defines the entities that are being counted (and distinguishes them from those other entities that are not to be counted). The other standard serves to determine the number of entities that are counted: this standard is the smallest subset of the positive integers that includes 1 and all of its successors (2, 3, ...) that can be put in one-to-one correspondence with the entities being counted: the largest integer in this set is the value of the count.

The following examples illustrate evaluations of the uncertainty associated with counts, and also very different ways of counting: leukocytes in a patient's blood smear (Page 38); woodlarks in Finland (Page 41); atoms of radon in air (Page 46); *Tyrannosaurus rex* before they became extinct (Page 49); and tramcars circulating in a city (Page 52).

These examples will also show that, as with other kinds of measurement, a count qualified with uncertainty can be used to inform an action or decision: whether some remediation is warranted to reduce the concentration of radon in a home, or whether some therapy is required to address a shortage of white blood cells in a patient's blood.

Counting Leukocytes

Leukocytes (white blood cells) are an important part of the immune system as they help fight infections by attacking bacteria, viruses, and other germs that invade the body. Thus, leukocyte count is commonly performed to detect hidden infections within the body. Fuentes-Arderiu and Dot-Bach [2009] report results of classifying and counting leukocytes of different types in a blood smear, known as a differential leukocyte count. The typical procedure when such counting is done manually while examining the sample under the microscope, is to count 100 leukocytes in total, while keeping a tally of the different types of leukocytes.

LEUKOCYTES	n	$u_S(n)$	$u_B(n)$
Neutrophils	63	5	4
Lymphocytes	18	4	6
Monocytes	8	3	4
Eosinophils	4	2	3
Basophils	1	1	3
Myelocytes	1	1	1
Metamyelocytes	5	2	4

Leukocyte count (n), uncertainty attributable to sampling variability ($u_S(n)$), and uncertainty attributable to differences between examiners ($u_B(n)$).

In this case, 4 eosinophils were counted among the 100 leukocytes. It is to be expected that, if another blood smear from the same patient were to be similarly examined, the number of eosinophils would turn out different from 4, owing to the vagaries of sampling.

This source of uncertainty is often modeled using either the binomial (Page 173) or the Poisson (Page 173) probability distributions. Since the probability of finding an eosinophil is small, these two models lead essentially to the same evaluation of this uncertainty component: that the proportion of eosinophils should vary by about $\sqrt{4}/100 = 2\,\%$ around the measured value of 4, which is taken as the estimate of the Poisson mean, whence the count will have standard deviation $\sqrt{4}$.

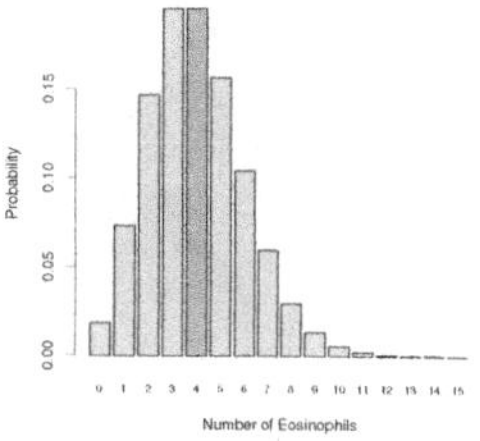

Probabilities from the Poisson distribution with mean 4 for the number of eosinophils in the differential leukocyte count listed above.

Counting the eosinophils involves: (i) identifying them, that is, defining the subset of the 100 leukocytes under examination that are eosinophils; (ii) actually counting the eosinophils that were identified; and (iii) qualifying the count with an evaluation of uncertainty, which should include contributions from sampling variability and from differences between examiners (which express identification uncertainty).

Holotype of a female *Agrias amydon phalcidon* butterfly from Brazil — Wikimedia Commons (Notafly, 2011).

The standard for the identification task (i) should be the *holotype* (paradigm, reference exemplar) for an eosinophil.

For species of plants and animals, the holotype is the individual specimen used to define a species, but there are no formal holotypes for different types of leukocytes. Because eosinophils are not identical copies of one another, accurate identification requires familiarity with their natural variability and reliance on distinctive traits that allow distinguishing them from the other types of leukocytes. For this reason, when different examiners count the same set of 100 leukocytes, it is likely that they will arrive at different counts for the different types of leukocytes.

Fuentes-Arderiu et al. [2007] have evaluated this source of uncertainty that is attributable to the effect of examiners, concluding that the coefficient of variation for the proportion of eosinophils was 69 %. Therefore, the uncertainty component for the count of eosinophils that arises from differences between examiners amounts to $4 \times 69\,\% = 3$ eosinophils.

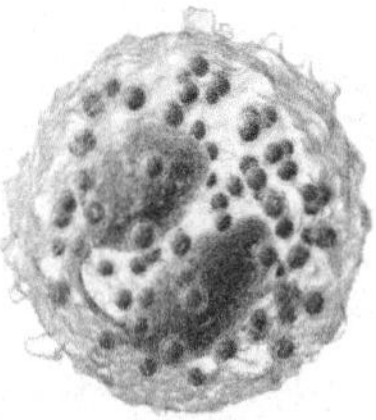

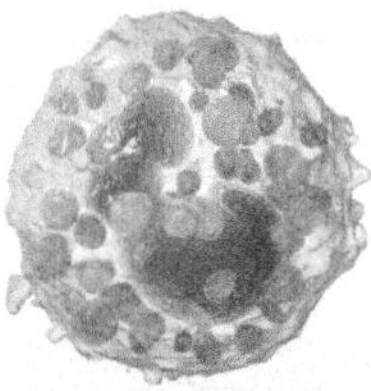

Eosinophils (TOP) are leukocytes that fight parasitic infections and mediate allergic reactions. Basophils (BOTTOM) control the response to allergens — Wikimedia Commons (BruceBlaus, 2017). Unless the blood smear being measured is stained to emphasize basophils, they may be confused with eosinophils.

The standard for the counting task (ii) is the unique finite set I comprising consecutive, positive integers, starting with 1, that can be put in one-to-one correspondence with the leukocytes that have been identified as being eosinophils: the measured value of the number of eosinophils is the largest integer in I. Task (ii) is counting *sensu stricto*, after identification, and is susceptible to counting errors. However, and in this case, since the numbers of leukocytes of the different types all are fairly small, and typically they are tallied using mechanical counters, we will assume that there are no counting errors above and beyond any identification errors.

Regarding task (iii), uncertainty evaluation, we need to take into account the fact that the total number of leukocytes that are identified and counted is fixed. Therefore, and for example, if an eosinophil is misclassified as a basophil, then the undercount for eosinophils results in an overcount for basophils.

This means that the uncertainty evaluation for the counts

cannot be performed separately for the different types of leukocytes, but must be done for all jointly, taking the effect of the fixed total into account: the so-called *closure constraint*.[26]

Performing a differential leukocyte count is equivalent to placing 100 balls (representing the 100 leukocytes) into 7 bins (representing the different types of leukocytes considered in this case), where the probability of a ball landing in a particular box is equal to the true proportion of the corresponding type of leukocyte in the subject's blood.

[26] F. Chayes. On correlation between variables of constant sum. *Journal of Geophysical Research*, 65(12):4185–4193, 1960. doi:10.1029/JZ065i012p04185

The probability model often used to describe the uncertainty associated with the numbers of balls that actually end-up in the different bins is the multinomial probability distribution (Page 176). This model also takes into account the fact that no count can be negative. For the eosinophils, considering both sampling and examiner sources of uncertainty, their true count is believed to lie between 0 and 10 with 95 % probability, using methods reviewed under *Counts* (Page 177).

Counting Woodlarks

The woodlark: lithograph by Magnus von Wright (1805-1868) — Wikimedia Commons

Each fall, the northern European woodlark (*Lullula arborea*) migrate south. During the 2009 fall migration season, woodlarks were counted at the Hanko bird observatory in southwestern Finland, by the Baltic Sea. These counts were made from the 245th through the 315th day of that year (September-November) [Lindén and Mäntyniemi, 2011], where n_k denotes the number of days with k sightings:

k	0	1	2	3	4	5	6	8	9	17	19	21	25	39
n_k	39	8	4	4	3	2	2	2	2	1	1	1	1	1

For example, there were 39 days with no sightings of woodlarks and there was one day where eight woodlarks were sighted. On average, there were 3.1 sightings each day, and the variance of the number of daily sightings was 44.

Both the Poisson distribution (Page 173) and the negative binomial distribution (Page 175) are common candidates for statistical modeling of counts.

The Poisson model may be appropriate for counts whose mean and variance are approximately equal, which is clearly not the case here.

The negative binomial distribution with mean μ and dispersion parameter ϕ can model counts that are more dispersed than Poisson counts, and lends itself to the following interpretation: the number of woodlarks sighted each day is like an outcome of a Poisson random variable, but the means of the daily counts vary and are like a sample from a gamma distribution with shape ϕ and scale μ/ϕ.[27]

[27] N. L. Johnson, A. W. Kemp, and S. Kotz. *Univariate Discrete Distributions*. John Wiley & Sons, Hoboken, NJ, Third edition, 2005. ISBN 0-471-27246-9

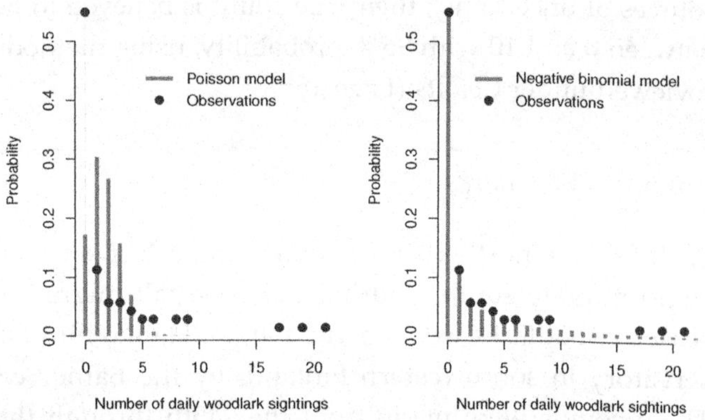

Probabilities from the negative binomial distribution with mean $\mu = 3.1$ and dispersion $\phi = 0.22$, and from the Poisson distribution with mean $\lambda = \mu$, along with observed relative frequencies of woodlark sightings.

The negative binomial model, calibrated with maximum likelihood estimates (Page 191) of the parameters, $\widehat{\mu} = 3.1(8)$ and $\widehat{\phi} = 0.22(5)$, fits these data quite well, and significantly better than the corresponding Poisson model.

The following R code implements maximum likelihood estimation (of the parameters of the negative binomial model) using R function `mle2` defined in package `bbmle`.[28]

[28] B. Bolker and R Development Core Team. *bbmle: Tools for General Maximum Likelihood Estimation*, 2020. URL `https://CRAN.R-project.org/package=bbmle`. R package version 1.0.23.1

```
k = c(0, 1, 2, 3, 4, 5, 6, 8, 9, 17, 19, 21, 25, 39)
nk = c(39, 8, 4, 4, 3, 2, 2, 2, 2, 1, 1, 1, 1, 1)
x = rep(k, times=nk)
a = mean(x)
v = var(x)

require(bbmle)
negloglik.negbin = function (mu, phi, x) {
  if ((mu < 0) | (phi < 0)) { return(Inf) }
  else { return(-1*sum(dnbinom(x, size=phi, mu=mu, log=TRUE))) }
  }
NB.mle = mle2(negloglik.negbin, start=list(mu=a, phi=a^2/(v-a)),
              skip.hessian=FALSE, method="Nelder-Mead",
              data=list(x=x))
summary(NB.mle)
```

The standard uncertainties associated with these parameters are the conventional large-sample approximations for MLEs. A more thorough uncertainty evaluation can be obtained by incorporating the sampling uncertainty. This can be done using a Monte Carlo method, as illustrated for maximum likelihood estimation of the Weibull distribution parameters (Page 193).

The large excess variance relative to the Poisson model, of $\widehat{\mu}/\widehat{\phi} = 137\,\%$, quantifies the extent of the overdispersion, which may be a consequence of the woodlarks' tendency to flock in small groups during autumn.

An alternative, Bayesian approach delivers estimates of the parameters, the evaluation of the associated uncertainties, and can also make predictions for future counts. The approach can be implemented in several different ways, and can incorporate varying amounts of prior knowledge about the flocks of visiting woodlarks.

The Bayesian model we shall consider expresses the parameters of the negative binomial distribution as functions of two hidden parameters α and β,

$$\mu = \alpha/\beta \quad \text{and} \quad \phi = \alpha,$$

and assigns half-Cauchy (Page 170) prior distributions to these hidden parameters. Below is the implementation of the Bayesian model in the Stan language.

```
// Negative binomial Stan model for woodlark sightings
nb_stan = "
data {
      int<lower=1> n;
      int<lower=0> x[n];
      real<lower=0> alphaM;
      real<lower=0> betaM;
      }
parameters {
      real<lower=0> alpha;
      real<lower=0> beta;
      }
transformed parameters {
      real mu = alpha/beta;
      real phi = alpha;
      }
model {
      // Half-Cauchy priors for alpha and beta
      // with medians alphaM and betaM
      alpha ~ cauchy(0, alphaM);
      beta ~ cauchy(0, betaM);
      // Likelihood
      x ~ neg_binomial(alpha, beta);
      }
  "
```

The following R code fits the Bayesian model defined above using facilities available in R package `rstan`.[29]

[29] B. Carpenter, A. Gelman, M. Hoffman, D. Lee, B. Goodrich, M. Betancourt, M. Brubaker, J. Guo, P. Li, and A. Riddell. Stan: A probabilistic programming language. *Journal of Statistical Software*, 76(1):1–32, 2017. doi:10.18637/jss.v076.i01; and Stan Development Team. *Stan User's Guide*. `mc-stan.org`, 2019. Stan Version 2.28

```
## Woodlark sighting data
k = c(0, 1, 2, 3, 4, 5, 6, 8, 9, 17, 19, 21, 25, 39)
nk = c(39, 8, 4, 4, 3, 2, 2, 2, 2, 1, 1, 1, 1, 1)
x = rep(k, times=nk)

## Fit Stan model
require(rstan)
nb.stan.fit = stan(model_code=nb_stan,
                   data=list(n=length(x), x=x,
                   alphaM=1, betaM=1))
print(nb.stan.fit)
```

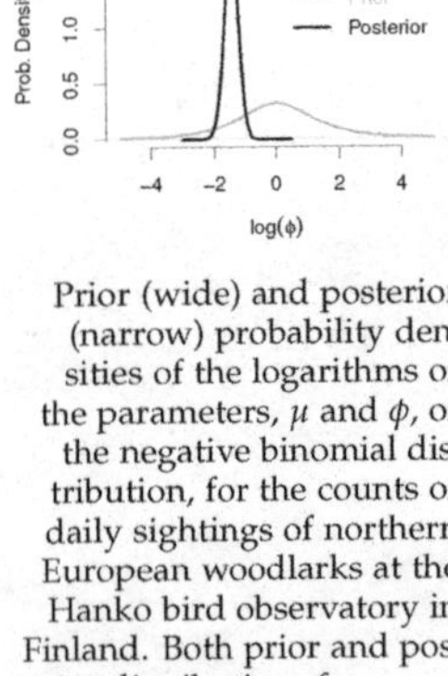

Prior (wide) and posterior (narrow) probability densities of the logarithms of the parameters, μ and ϕ, of the negative binomial distribution, for the counts of daily sightings of northern European woodlarks at the Hanko bird observatory in Finland. Both prior and posterior distributions for μ and ϕ are markedly skewed to the right, but the corresponding distributions of their logarithms, depicted here, are approximately symmetrical.

The prior probability distributions for α and β both have median 1, hence express the belief that both μ and ϕ are equally likely to be smaller than 1 or larger than 1 a priori.

The posterior distribution of μ has mean 3.1 and standard deviation 0.8, and the posterior distribution of ϕ has mean 0.24 and standard deviation 0.06. These estimates are nearly identical to those provided either by the maximum likelihood or Monte Carlo methods.

METHOD	$\widehat{\mu}$	$\widehat{\phi}$
Maximum Likelihood	3.1(8)	0.22(5)
Monte Carlo	3.1(8)	0.23(6)
Bayes	3.1(8)	0.24(6)

The *predictive distribution* is the conditional probability distribution of a future daily count, y, given the daily counts, $x_1, \dots, x_n$, that were observed on $n = 71$ different days. This distribution is meaningful provided the migration process remains stable during the migration period.

Even though the formal calculation of the corresponding predictive distribution is rather forbidding, it is easy to draw a sample from the predictive distribution, which suffices for all practical purposes. These are the steps to generate each value in the sample:

- Draw a pair of values of the parameters, μ^* and ϕ^*, from the MCMC sample drawn previously from their joint posterior distribution;
- Draw a value from a negative binominal distribution with mean μ^* and dispersion parameter ϕ^*.

The following R code yields $y = [0,\ 22]$ as a 95 % predictive interval, meaning that, with 95 % probability, a future daily count of woodlarks should not exceed 22.

```
mu.post = extract(nb.stan.fit)$mu
phi.post = extract(nb.stan.fit)$phi
K = length(mu.post)
y = rnbinom(n=K, size=phi.post, mu=mu.post)
round(quantile(y, probs=c((1-0.95)/2, (1+0.95)/2)), 2)
```

This prediction, of course, pertains to the conditional distribution of a future observation, y, given the observations already made.

Counting Radon Atoms

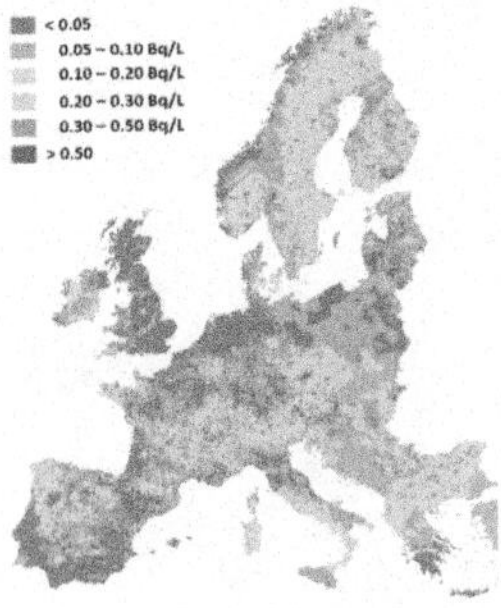

European map of indoor radon.
J. Elío, G. Cinelli, P. Bossew, J. L. Gutiérrez-Villanueva, T. Tollefsen, M. De Cort, A. Nogarotto, and R. Braga. The first version of the Pan-European Indoor Radon Map. *Natural Hazards and Earth System Sciences*, 19(11):2451–2464, 2019. doi:10.5194/nhess-19-2451-2019

Radon is a colorless, odorless, tasteless, radioactive gas produced naturally by radioactive decay of uranium, which occurs in many rocks and soils, particularly in regions where granite is common.

The World Health Organization (WHO) recommends that homeowners should take remedial action if the activity of radon exceeds 0.100 Bq/L of the air inside their homes [Zeeb and Shannoun, 2009].

The half-life of ^{222}Rn is $t_{½} = 3.8215$ days [Kondev et al., 2021], and each of its atoms decays by emission of an α-particle into an atom of ^{218}Po, which is also radioactive. If radon is inhaled, the α-particles it emits will damage the lungs. The WHO estimates that radon causes up to 250 000 deaths from lung cancer each year, worldwide.

Those 0.100 Bq/L correspond to $N = 47\,635$ atoms of ^{222}Rn per liter of air. We assume that this number concentration remains approximately constant over time as decaying atoms of radon are replaced by fresh radon that continuously seeps into the home through cracks and openings in its foundation, from the surrounding soil and rocks. Of those many atoms, about 1 will decay every 10 seconds.

There are $N_0 = N_A / M(^{222}\text{Rn}) \approx 2.71 \times 10^{21}$ atoms in 1 gram of ^{222}Rn, of which $N_1 = N_0(1 - \exp(-1/\tau)) \approx 5.69 \times 10^{15}$ are expected to decay within one second. Here, $N_A = 6.022\,140\,76 \times 10^{23}$ mol is the Avogadro constant, $M(^{222}\text{Rn}) = 222.017\,576$ g/mol is the molar mass of ^{222}Rn, and $\tau = t_{½}/\ln(2) = 132.3$ h is the expected lifetime of each atom. Since 1 Bq means one radioactive decay per second, it follows that 0.100 Bq/L corresponds to there being $N = (N_0/N_1) \times (0.100\text{ Bq/L}) = 47\,635$ atoms per liter.

Even though all atoms of ^{222}Rn are identical, they decay at unpredictable and different times, each independently of the others. The lifetime of a radionuclide that has a single mode of decay (which is emission of an α-particle in the case of radon), is like an outcome of a random variable with an exponential distribution (Page 171) whose median is the half-life. This distribution captures the fact that a radionuclide does not age. In other words, it does not "remember" when it was born.

The lifetimes of the different atoms of radon in the assembly of N atoms mentioned above are mutually independent, exponential random variables, hence all enjoy the lack of memory just mentioned.

Fix a particular instant in time after which we await the emission of the first α-particle from this assembly: it will originate from the atom whose actual lifetime is the shortest of the lifetimes of the N atoms under observation.

Now, the shortest of N mutually independent, exponentially distributed lifetimes, all of which have the same expected lifetime τ, also has an exponential distribution, but its expected value is τ/N.

If $L_1, \ldots L_N$ denote N mutually independent, exponentially distributed (Page 171) lifetimes and M denotes the shortest among them, then $\Pr(M \leqslant m) = 1 - \Pr(M > m) = \Pr(L_1 > m, \ldots, L_N > m) = \Pr(L_1 > m) \ldots \Pr(L_N > m) = 1 - \exp(-m/(\tau/N))$ owing to independence and because $\Pr(L_j > m) = \exp(-m/\tau)$ for $j = 1, \ldots, N$.

Since we have assumed that the assembly of N atoms is constantly replenished, the waiting time between the first and second emissions will have the same distribution as the waiting time until the first emission, and the same for the waiting time between the second and third, and so on.

Now, exponential waiting times between consecutive emissions imply that the "arrivals" of consecutive α-particles issuing from the steady-state assembly of N radon atoms are a realization of a so-called *Poisson process* with rate N/τ [Grimmett and Welsh, 2014, Theorem 11.3], where $\tau = t_{½}/\ln(2)$ is the expected lifetime of each atom. This, in turn, implies that the number of α-particles emitted during a specified time interval is like an outcome of a Poisson (Page 173) random variable whose mean depends on the original number, N, of atoms, on their lifetimes, τ, and on the duration of the interval.

In 1910, Ernest Rutherford and Hans Geiger [Rutherford et al., 1910], confirmed these facts empirically by timing and counting a large number of scintillations produced by α-particles emanating from a sample of polonium as they hit a zinc sulfide screen.

A note prepared by Harry Bateman, which was added to Rutherford and Geiger's article, explains why the counts should be like a sample from a Poisson distribution provided the rate of scintillations remains constant in the course of the experiment. This was achieved by moving the polonium source "daily closer to the screen," thus correcting for the steadily decreasing number of radioisotopes in the source.

In the case of radon in a home's basement, the decays are a homogeneous Poisson process (Page 173) because, as pointed out already, decaying atoms are constantly being replaced by new atoms seeping into the house through cracks and openings in its foundation.

The radioactive decay of radon atoms (or any other radionuclide that emits α-particles) is nothing short of a miracle because it entails an α-particle overcoming a barrier of much greater energy than the particle possesses. The fact that an α-particle has the wherewithal to leave the nucleus of an atom of radon is as surprising as it would be for a ping-pong ball to pass through the playing table intact instead of bouncing off it. Such miraculous (quantum) "tunneling" does not happen with ping-pong balls, but it can and does happen with

[30] R. W. Gurney and E. U. Condon. Wave mechanics and radioactive disintegration. *Nature*, 122:439, September 1928. doi:10.1038/122439a0

α-particles. And the manner in which it happens does not involve brute force, rather "the α-particle slips away almost unnoticed."[30]

The pattern of decay of radon has these noteworthy characteristics, where $\lambda = 1/\tau$ is the *decay rate*:

- The *expected* number $N\exp(-\lambda t)$ of atoms of radon remaining, out of the original $N = 47\,635$ atoms per liter of air, decays exponentially fast with time t;
- The *actual* number of atoms that will decay in the time period from t_1 to t_2 ($t_1 > 0$ and $t_2 > t_1$) is like an outcome of a Poisson random variable with mean $N(\exp(-\lambda t_1) - \exp(-\lambda t_2))$.

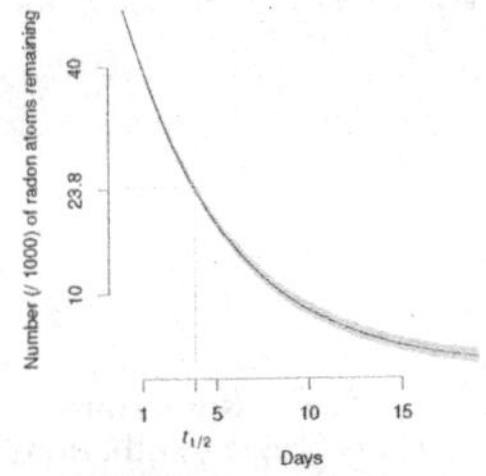

Expected number, out of the original 47 635 atoms of radon per liter of air, that remain after collecting the sample on day 0 (dark line), and surrounding band where the actual numbers should lie with 95 % probability.

MEASURING VOLUMIC ACTIVITY: A sample of 1 L of air was collected from the basement of a home, and placed in a hermetically sealed container opaque to α-particles. The sample yielded $D = 693$ α-particles during the first hour after collection. The corresponding volumic activity of radon in the basement, defined as the steady-state number of α-particles emitted per second and per liter of air, is

$$A_V = \frac{D}{1\,\text{L}} \frac{1 - \exp\left\{-\dfrac{\ln(2)}{t_{1/2} \times 86400\ \text{s/d}}\right\}}{1 - \exp\left\{-\dfrac{\ln(2)}{t_{1/2} \times 24\ \text{h/d}}\right\}},$$

where D, the number of atoms of radon that decayed during the first hour after collecting the sample, is modeled as a Poisson random variable with mean 693, and the half-life of ^{222}Rn, $t_{1/2}$, is modeled as a Gaussian random variable with mean 3.8215 days and standard deviation 0.0002 days [Kondev et al., 2021].

Both Gauss's formula (Equation (10) in the GUM) and the Monte Carlo method implemented in the *NIST Uncertainty Machine*, produce $A_V = 0.193\,\text{Bq/L}$ with standard uncertainty $u(A_V) = 0.007\,\text{Bq/L}$, thus suggesting the need for remediation.

Counting Dinosaurs

Tyrannosaurus rex, one of the largest predators ever to live on land, roamed western North America some 70 million years ago, during a period that lasted several million years and ended abruptly when an asteroid hit the Yucatán Peninsula, in the Gulf of Mexico [Renne et al., 2013].

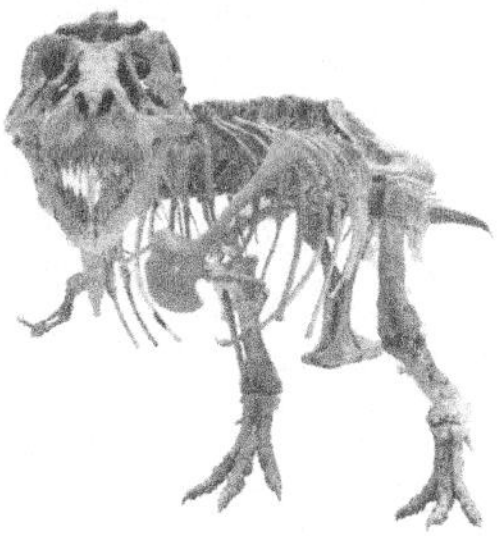

Replica of SUE, one of the most complete specimens of *Tyrannosaurus rex* ever found, displayed in the Field Museum of Natural History in Chicago. (Chase Elliott Clark, Wikimedia Commons)

This *T. rex*, whose gender is unknown, was named after Sue Hendrickson, who discovered it in 1990.

It is estimated that at any particular time during that period there would have been some 20 000 *T. rex* roaming throughout their habitat [Marshall et al., 2021].

How is it possible to estimate the size of the population of *T. rex*? The approach described next relies on estimating two quantities whose product provides the answer: their population density and the area of their habitat.

The plot alongside depicts the relationship between population density and adult body mass for a large collection of predominantly terrestrial mammals, ranging from the minute Etruscan shrew to the colossal African bush elephant. The straight line has equation

$$\log_{10}(D/\mathrm{km}^{-2}) = 3.9 - 0.75 \times \log_{10}(M/\mathrm{kg}),$$

where M denotes body mass and D denotes population density. The slope, -0.75 with standard uncertainty 0.02, is consistent with *Damuth's Law* [Damuth, 1987, 2007], which applies to populations of many different land animals, and is one of the key elements in the production of the virtual count of *T. rex*.

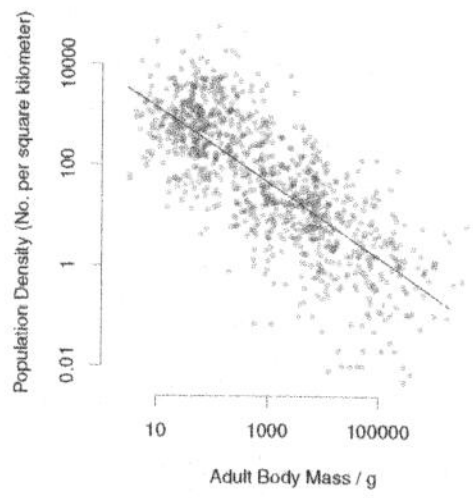

Relationship between population density and adult body mass for 5327 mammals from the PanTHERIA database [Jones et al., 2009], excluding marine mammals.

While the slope applies quite widely, the intercept varies for different taxonomic units, depending both on where in the food chain the animals are situated (trophic level) and on their physiology.

Marshall et al. [2021] argue that the uncertainty surrounding the intercept is very unlikely to exceed the variability of the intercept when Damuth's Law is fitted only to data pertaining to mammals that are herbivores or to mammals that are carnivores, and therefore assume that the intercept should be between 1.80 and 4.18 with 95 % confidence. Accordingly, we will model this state of knowledge about the value of this intercept for *T. rex* using a Gaussian distribution with mean $(1.80 + 4.18)/2 = 2.99$ (which differs from the intercept derived for the mammals listed in the PanTHERIA database) and standard deviation 0.61.

Additionally, we will use the estimate of the slope derived from the PanTHERIA database, and model its associated uncertainty using a Gaussian distribution with mean -0.75 and standard deviation 0.02.

To derive an estimate the population density of *T. rex* from Damuth's law, we need to know their body mass. But how does one weigh a dinosaur? Similarly to Damuth's law, we can take advantage of a universal scaling relationship between the body mass and the minimum circumference of the femur bone for animals that are alive today.[31]

[31] N. Campione and D. C. Evans. A universal scaling relationship between body mass and proximal limb bone dimensions in quadrupedal terrestrial tetrapods. *BMC Biology*, 10:60, 2012. doi:10.1186/1741-7007-10-60; and N. E. Campione and D. C. Evans. The accuracy and precision of body mass estimation in non-avian dinosaurs. *Biological Reviews*, 95(6):1759–1797, 2020. doi:10.1111/brv.12638

For the adult body mass of *T. rex*, Marshall et al. [2021] believe that its 2.5th and 97.5th percentiles should have been 3700 kg and 6900 kg, with typical value 5200 kg, thus comparable to the mass of the African forest elephant, of about 4500 kg. Since this interval is asymmetric about the typical value, we will model this uncertainty using a skew-normal distribution[32] fitted to these data as described by Possolo et al. [2019]. The following R code yields location $\xi = 4608$ kg, scale $\omega = 1023$ kg, and shape $\alpha = 1.16$ for this distribution.

[32] A. Azzalini and A. Capitanio. *The Skew-Normal and Related Families*. Cambridge University Press, Cambridge, UK, 2014. ISBN 978-1-107-02927-9. doi:10.1017/cbo9781139248891

```
require(sn)
q = c(0.025, 0.500, 0.975)
x = c(3700, 5200, 6900)
msn = function(p) {
  q = qsn(q, xi=p[1], omega=p[2], alpha=p[3])
  sum((q - x)^2)
  }
optim(c(5000, 1000, 2), msn, lower=c(0,0,-10),
      upper=c(Inf, Inf, 10), method='L-BFGS-B')
```

Finally, Marshall et al. [2021] estimate the area of the geographic range for *T. rex* as $A = (2.3 \pm 0.88) \times 10^6\,\text{km}^2$, with 95 % confidence. We will model this uncertainty using a lognormal distribution (Page 172) with mean $2.3 \times 10^6\,\text{km}^2$ and standard deviation $0.44 \times 10^6\,\text{km}^2$.

The following R code employs a Monte Carlo method for uncertainty propagation, and also produces an estimate of the population density of *T. rex* (median of 0.009 individuals per square kilometer), and a virtual count of the number of *T. rex* roaming their range at any particular time: this count has median 20 000, and a 95 % coverage interval for it ranges from 1200 to 360 000.

```
K = 1e7

## Intercept for Damuth's Law
a = rnorm(K, mean=2.99, sd=0.61)

## Slope for Damuth's Law
b = rnorm(K, mean=-0.75, sd=0.021)

## Skew-normal model for adult body mass
library(sn)
M = 1000*rsn(K, xi=4608, omega=1023, alpha=1.16)

## Population density (number of T. rex per km^2)
D = 10^(a + b*log10(M))

## Lognormal model for area of geographical range
mu = 2.3; sigma = 0.88/2
mulog = log(mu/sqrt((sigma/mu)^2 + 1))
sigmalog = sqrt(log((sigma/mu)^2 + 1))
A = rlnorm(K, mean=mulog, sd=sigmalog)*1e6

## Median and coverage interval for number of individuals
signif(c(median(A*D), quantile(A*D, probs=c(0.025, 0.975))), 2)
```

Counting Tramcars

For most of the 20th century, tramcars were a reliable means of public transportation in many cities, like this tramcar serving route 28 in Lisbon, Portugal (Photo by Oriol Pascual, 2019, on Unsplash).

> "A man travelling in a foreign country has to change trains at a junction, and goes into the town, of the existence of which he has only just heard. He has no idea of its size. The first thing that he sees is a tramcar numbered 100. What can he infer about the number of tramcars in the town? It may be assumed for the purpose that they are numbered consecutively from 1 upwards." — Harold Jeffreys (1939, §4.8)

Harold Jeffreys credits Max Newman with having suggested this problem to him in the 1930s, when both were fellows of St. John's College, Cambridge. Roy Geary, writing in 1944, reported that "at a recent meeting of the Dublin University Mathematical Society," E. Schrödinger asked a more general question: given the serial numbers of n cars known to be numbered sequentially from 1 to an unknown number M, estimate this number.[33]

The same question was asked about Germany's rate of production of *Panther* tanks during February 1944, as part of the intelligence gathering intended to support the D-day invasion of Normandy.[34]

The question was answered based on the analysis of bogie wheel markings from two such tanks — one captured in Russia in March of 1943, the other captured in Sicily in February of 1944 — which allowed estimating the number of molds being used to produce such wheels. Coupled with expert knowledge about the number of times such molds could be reused, that analysis led to an estimate of 270 tanks having been produced in February of 1944. The actual number, determined much later based on German production records, had been 276. Similar successes were achieved for other tank models produced during other periods of World War II.

In the case of tramcars, the measurand is the number M of tramcars in the town, which are assumed to be equally likely to be sighted. The classical estimate of M, given observations $m_1, \dots, m_n$ of the serial numbers

[33] R. Geary. Comparison of the concepts of efficiency and closeness for consistent estimates of a parameter. *Biometrika*, 33:123–128, 1944. doi:10.2307/2334111

[34] R. Ruggles and H. Brodie. An empirical approach to economic intelligence in World War II. *Journal of the American Statistical Association*, 42(237):72–91, 1947. doi:10.2307/2280189

of n tramcars, is $\widehat{M} = (1 + 1/n)m^* - 1$, where $m^* = \max\{m_1, \ldots, m_n\}$.[35] In the version of the problem that Jeffreys considered, for which $n = 1$ and $m = 100$, this estimate is $\widehat{M} = 199$.

[35] G. Clark, A. Gonye, and S. J. Miller. Lessons from the German Tank Problem. *arXiv e-prints*, page arXiv:2101.08162 [stat.OT], 2021. URL https://arxiv.org/abs/1905.12362

Next we review Jeffreys's approach, which offers a Bayesian explanation for the intuitive "feeling that there is something special about the value $2m$." The probability of first sighting tramcar number m is $p(m|M) = \{m \leqslant M\}/M$, where the expression in the numerator, $\{m \leqslant M\}$, stands for an indicator function, being 1 when the condition between the curly brackets is satisfied, and 0 otherwise. With $m = 100$ fixed at the number of the first tramcar that was sighted, the value of M that maximizes $p(m|M)$ is $\widehat{M} = m$ because $p(m|M)$ will be 0 unless $M \geqslant m$, and $1/M$ decreases with increasing M. That is, $m = 100$ is the maximum likelihood estimate (Page 191) of M, which is about half the size of the classical estimate and just does not seem reasonable.

If $x_1, \ldots, x_n$ are a sample from a probability distribution whose density (Page 159) is p_θ, where θ denotes a (possibly vectorial) parameter, then the maximum likelihood estimate of θ based on this sample is the value $\widehat{\theta}$ that maximizes the product $p_\theta(x_1) \ldots p_\theta(x_n)$ with respect to θ, while the $\{x_j\}$ remain fixed at their observed values.

Jeffreys's Bayesian approach requires that M also be assigned a probability distribution, whose role is to capture the visitor's complete ignorance about the true value of M. A uniform (or, rectangular, Page 167) distribution is often used for such purpose: for example, the prior probability is $1/6$ that a casino die, when rolled, will come to rest with M pips facing up, for $M = 1, \ldots, 6$. This probability distribution expresses the prior belief that the die is a perfectly balanced cube and that the roll will be sufficiently chaotic to make the outcome unpredictable. But Jeffreys assigned a different prior distribution to M arguing that it is inappropriate to use the uniform distribution as "a way of saying that the magnitude of a parameter is unknown" when the parameter is positive and can be arbitrarily large.

Jeffreys's choice was driven by an invariance argument [Jeffreys, 1961, §3.10] suggesting that the "right" prior distribution for a non-negative, otherwise unrestricted, positive parameter M should have a probability density proportional to $1/M$. Jeffreys took one additional step and suggested that it should be so also when the parameter of interest is a positive integer.

The problem is that it is not possible to assign probabilities proportional to $1/M$ when the range of M comprises all the positive integers because the required constant of proportionality would be $1/1 + 1/2 + 1/3 + \ldots$, which is infinity.

In his original treatment, consistently with his theory of invariant prior distributions, Jeffreys did not introduce a finite, even if very large, upper bound for M. Instead, he used an unrealistic but practicable mathematical device called an *improper prior distribution*, which, in this case, involves defining a prior probability mass function (Page 160) p such that $p(M) = c/M$ for $M = 1, 2, \ldots$, and leaving the "normalizing" constant c unspecified. The device works because, once this prior mass function is used in Bayes's rule, the constant c appears in both numerator and denominator, hence cancels, and the posterior mass function becomes

$$q(M|m) = \frac{\{M \geqslant m\}/M^2}{\sum_{k=0}^{\infty} 1/(m+k)^2}.$$

The denominator is the value $\zeta(2, m)$ of the Hurwitz zeta function [Apostol, 2010], which can be evaluated numerically using R function `zeta` defined in package `VGAM` [Yee, 2010], thus allowing the exact calculation of the quantiles of the posterior distribution, which are practically identical to those obtained using the approximation and R code listed alongside.

One can avoid this problem by imposing a large yet finite upper limit for the value of M, say a value that is million times larger than the observed m. Combining this prior distribution with the likelihood function via Bayes's rule, one obtains a posterior distribution with probability density

$$q(M|m) = \frac{\{M \geqslant m\}/M^2}{\sum_{k=m}^{10^6 m} 1/k^2}, \quad \text{for } M = 1, 2, \ldots$$

The following R code, where `D` denotes the value of the denominator in the above formula for $q(M|m)$, computes the median of this posterior distribution, and the endpoints of a 95 % credible interval:

```
m = 100
D = sum(1/seq(from=m, to=1e6*m)^2)
x = seq(from=m, to=1e6*m)
qM = (1/x^2)/D

x[which.min(abs(cumsum(qM)-0.5))]
x[which.min(abs(cumsum(qM)-0.025))]
x[which.min(abs(cumsum(qM)-0.975))]
```

The posterior median of M is 199, thus justifying the intuitive inclination toward the answer to Jeffreys's problem being $M = 2m$. More strikingly, the 95 % credible interval extends from 102 to 3979 and does not even include the observed value of the measurand!

The more general problem, where the serial numbers $m_1, \ldots, m_n$ of n tramcars have been observed, can be solved noting that the likelihood function is

$$p(m_1,\ldots,m_n|M) = \{m_1 \leqslant M\} \times \cdots \times \{m_n \leqslant M\}/M^n$$
$$= \{m^* \leqslant M\}/M^n,$$

where $m^* = \max\{m_1,\ldots,m_n\}$. The posterior density now becomes

$$q(M|m_1,\ldots,m_n) = \frac{\{M \geqslant m^*\}/M^{n+1}}{\sum_{k=m^*}^{10^6 m} 1/k^{n+1}}, \text{ for } M = 1,2,\ldots$$

Since only the largest of the n serial numbers plays a role in $q(M|m_1,\ldots,m_n)$, one might think that the posterior median of M given the observations should be largely unaffected if the largest serial number remains the same. The following illustration proves this conjecture wrong, involving the observation of $n = 4$ serial numbers, $\{100, 71, 23, 89\}$, whose maximum is the same as before: $m^* = 100$.

```
m = c(100, 71, 23, 89)
mSTAR = max(m)
D = sum(1/seq(from=mSTAR, to=1e6*mSTAR)^(4+1))
x = seq(from=mSTAR, to=1e6*mSTAR)
qM = (1/x^(4+1))/D

x[which.min(abs(cumsum(qM)-0.5))]
x[which.min(abs(cumsum(qM)-0.025))]
x[which.min(abs(cumsum(qM)-0.975))]
```

The posterior median of M is only 118, and a 95 % credible interval for M ranges from 100 to 250. The reason why the posterior median is now so much smaller than when a single serial number had been sighted, even though the largest serial number is the same in both cases, is that the maximum of a large sample drawn from a uniform distribution is more likely to lie within a specified distance of the upper end-point of the interval where the distribution is concentrated, than the maximum of a small sample.

Surveying

In 2019, non-irrigated pastureland in Kansas was valued at around \$4620 per hectare ($1\,\text{ha} = 10\,000\,\text{m}^2$). A plot, shaped like an irregular heptagon on an essentially flat plain, is for sale with asking price \$206 000. The seller offered to provide coordinates of the vertices in triplicate, determined using a portable, consumer-grade GPS receiver.

Triplicate determinations of the coordinates of the vertices of a heptagonal plot of pastureland in Kansas, USA. One of these determinations yielded 826 m easting and 615 m northing for vertex A, following the same pattern for the other vertices.

	EASTING / m			NORTHING / m		
A	826	821	848	615	625	619
B	673	698	699	752	782	763
C	440	419	434	781	795	802
D	82	98	107	415	411	380
E	131	121	115	149	105	117
F	471	495	480	−9	42	14
G	796	807	777	217	258	225

The potential buyer insisted that the triplicates should be obtained in three separate surveys. In each survey, the vertices were visited in random order, and the GPS receiver was turned off after taking a reading at a vertex, and then turned on again upon arrival at the next vertex, so that it would reacquire satellites and determine the location afresh.

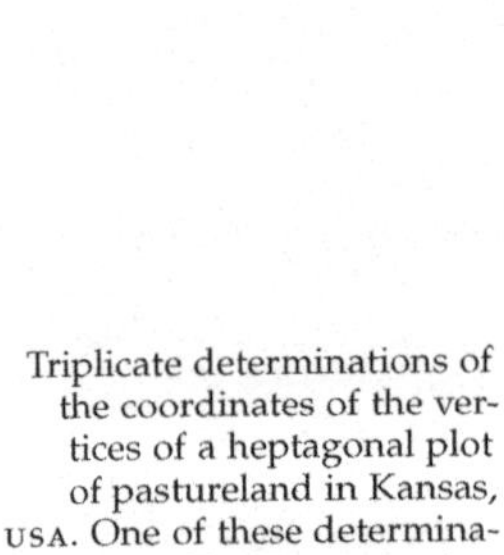

Plot of pastureland in Kansas, USA. The black dots mark the triplicates of the vertices as determined by a GPS receiver.

These are the questions the potential buyer wishes a surveyor will answer: (i) How to estimate the plot's area? (ii) How to evaluate the uncertainty surrounding this estimate? (iii) How may have the seller come up with that asking price? The reason for this last question is that some understanding of the origin of the asking price may be a valuable element when the potential buyer will make a decision about how much to offer.

To estimate the plot's area one may use the *Surveyor's Formula*.[36] However, before using it, one needs to decide how to combine the triplicate determinations of the location of each vertex. One way consists of averaging

[36] B. Braden. The surveyor's area formula. *The College Mathematics Journal*, 17(4):326–337, 1986. doi:10.2307/2686282

them. For example, the average easting for vertex A is

$$e(\mathrm{A})/\mathrm{m} = (826 + 821 + 848)/3 = 831.7.$$

Let $(e(\mathrm{A}), n(\mathrm{A}))$, $(e(\mathrm{B}), n(\mathrm{B}))$, ..., $(e(\mathrm{G}), n(\mathrm{G}))$ denote the averages of the Cartesian coordinates (easting and northing) of the triplicates at each vertex of the polygon in counterclockwise order (A, B, ..., G). These are the coordinates of the large (gray) dots in the plot alongside. The area of the shaded polygon is $S = 41.3\,\mathrm{ha}$, and it was computed as follows:

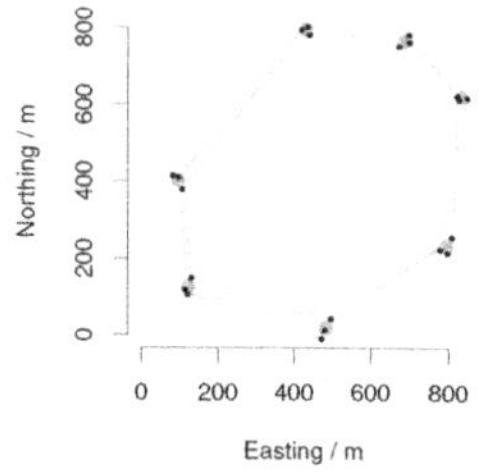

Four of the $3^7 = 2187$ heptagons that can be constructed using the replicate determinations of the vertices.

$$S = \frac{1}{2}\left(\begin{vmatrix} e(\mathrm{A}) & e(\mathrm{B}) \\ n(\mathrm{A}) & n(\mathrm{B}) \end{vmatrix} + \begin{vmatrix} e(\mathrm{B}) & e(\mathrm{C}) \\ n(\mathrm{B}) & n(\mathrm{C}) \end{vmatrix} + \cdots + \right.$$
$$\left. + \begin{vmatrix} e(\mathrm{F}) & e(\mathrm{G}) \\ n(\mathrm{F}) & n(\mathrm{G}) \end{vmatrix} + \begin{vmatrix} e(\mathrm{G}) & e(\mathrm{A}) \\ n(\mathrm{G}) & n(\mathrm{A}) \end{vmatrix} \right),$$

where $\begin{vmatrix} a & b \\ c & d \end{vmatrix} = ad - bc$.

The question may well be asked of why we used the averages of the triplicates, instead of some other summary. The average will be optimal when the measurement errors affecting the easting and northing coordinates are independent and Gaussian, and the goal is to minimize the mean squared error (Page 167) of the estimates of the vertices.

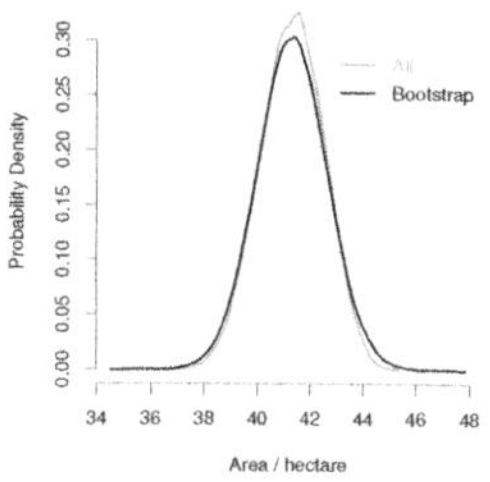

Probability density (Page 159) estimates for the area of the heptagon: based on the areas of all 2187 alternative heptagons, and on the parametric bootstrap (explained on Page 59). The former (gray, taller) ignores the correlations between the areas of the alternative polygons: the corresponding standard deviation is 1.17 ha. The latter (black, shorter) reflects the impact of measurement errors affecting the easting and northing coordinates of each vertex, and recognizes the small numbers of replicates per vertex: it has heavier tails, and the corresponding standard deviation is 1.32 ha.

Given the replicated determinations that were made of the locations of the vertices, it is possible to construct many different versions of the heptagon by choosing one of the three replicates made for vertex A, one of the three made for vertex B, etc. Each of these heptagons is consistent with the measurements that were made. Running through all $3^7 = 2187$ possible combinations of vertex determinations (each of which comprises a pair of values of easting and northing), and computing the areas of these alternative heptagons, yields a set of 2187 conceivable values for the area, whose average and median both equal 41.3 ha.

The area of the largest of these 2187 heptagons is 44.6 ha,

with corresponding value 44.6 ha × \$4620/ha ≈ \$206 000, which explains the likely rationale behind the asking price. Since the area of the smallest heptagon is 37.6 ha, the same rationale would support an offer of 37.6 ha × \$4620/ha ≈ \$174 000.

However, an offer based on a value for the area close to the average area is more likely to be accepted by the seller than one that is as deviant from the average on the low side, considering that the seller's asking price is based on the maximum area consistent with the measurements. But the buyer should also take into account the uncertainty associated with the area.

Considering that each replicate of each vertex appears in $3^6 = 729$ heptagons built as just described, hence that there are correlations between the 2187 areas of the alternative heptagons, the standard deviation of these areas, 1.17 ha, may not be a reliable evaluation of the uncertainty associated with the area of the plot of land.

```
east = array(c(826, 673, 440, 82, 131, 471, 796,
               821, 698, 419, 98, 121, 495, 807,
               848, 699, 434, 107, 115, 480, 777),
               dim=c(7,3))
north = array(c(615, 752, 781, 415, 149, -9, 217,
                625, 782, 795, 411, 105, 42, 258,
                619, 763, 802, 380, 117, 14, 225),
               dim=c(7,3))
z = data.frame(east=c(east), north=c(north),
          east.vertex=I(paste0("E", rep(1:7, 3))),
          north.vertex=I(paste0("N", rep(1:7, 3))))
```

To evaluate this uncertainty, the buyer hires a statistician, whose first task is to quantify the uncertainty associated with the measurement of each vertex. The statistician applies the Fligner-Killeen test[37] to the replicated determinations of the coordinates of the vertices of plot, and concludes that there is no reason to doubt that all 14 sets of replicates have the same variance.

[37] M. A. Fligner and T. J. Killeen. Distribution-free two-sample tests for scale. *Journal of the American Statistical Association*, 71(353):210–213, March 1976. doi:10.2307/2285771

The statistician proceeds by pooling the variances of the 14 groups of replicates, which yields a standard uncertainty of 16 m (on 28 degrees of freedom) for an individual determination of the easting or northing of a vertex.

```
fligner.test(x=c(z$east, z$north),
             g=c(z$east.vertex, z$north.vertex))

east.s = apply(east, 1, sd)
north.s = apply(north, 1, sd)
s = sqrt(sum((3-1)*east.s^2 + (3-1)*north.s^2) /
       ((3-1)*length(east.s) + (3-1)*length(north.s)))
s.nu = (3-1)*length(east.s) + (3-1)*length(north.s)
c(s=s, s.nu=s.nu)
```

The pooled variance for easting and northing is the sum of the sums of squared deviations from their averages for the values of easting and northing, over all the vertices, divided by the sum of the corresponding numbers of degrees of freedom $(3-1)$ per vertex. The pooled standard deviation, s, is the square root of the pooled variance.

The statistician's next task is to propagate this uncertainty to the uncertainty of the area, which she does employing the parametric statistical bootstrap [Efron and Tibshirani, 1993] (Page 179). This involves repeating the following two steps a large number of times:

- For each vertex $i = 1,\ldots,7$ in turn, simulate an easting of the form $e_i + \varepsilon_i$ and a northing of the form $n_i + \nu_i$, where (e_i, n_i) are the averages of the three determinations of easting and northing of vertex $i = 1,\ldots,7$, and ε_i and ν_i represent measurement errors with zero mean and standard deviation 16 m — these measurement errors are drawings from Student's t distributions with 28 degrees of freedom, rescaled to have this standard deviation.

- Use the Surveyor's Formula to compute the area of the heptagon whose vertices' locations were simulated in the previous step.

The statistician repeated these steps one million times and found that the average of the areas of the simulated heptagons was the same as the area determined originally, and that the standard deviation of the simulated areas was 1.3 ha. In light of this fact, the statistician suggested to the buyer than an offer between (41.3 ± 1.3)ha $\times$ \$4620/ha would be reasonable, that is between \$184 800 and \$198 612.

```
e = apply(east, 1, mean)
n = apply(north, 1, mean)
m = length(e)

K = 1e6
areaB = numeric(K)
for (k in 1:K)
{
  eB = e + s * rt(m, df=s.nu)/sqrt(s.nu/(s.nu-2))
  nB = n + s * rt(m, df=s.nu)/sqrt(s.nu/(s.nu-2))
  surv = (eB[m]*nB[1] - nB[m]*eB[1])
  for (i in 1:(m-1)) {
    surv = surv + (eB[i]*nB[i+I] - nB[i]*eB[i+1])}
  areaB[k] = (abs(surv)/2) / 10000
}
c(mean(areaB), sd(areaB))
```

The case just discussed involves a rather simple geometric figure: a heptagon whose boundary is clearly well defined. In practice, one often has to deal with more complex situations. Benoit Mandelbrot[38] famously asked the question "How long is the coast of Britain?" It so turns out that the answer to this question depends on the spatial scale at which the question is considered: or, in other words, on the size of the ruler used to measure it. Mandelbrot notes that "geographical curves are so involved in their detail that their lengths are often infinite or, rather, undefinable." In fact, the apparent length of the coastline decreases as the length of the ruler increases.

[38] B. Mandelbrot. How long is the coast of Britain? Statistical self-similarity and fractional dimension. *Science*, 156:636–638, May 1967. doi:10.1126/science.156.3775.636

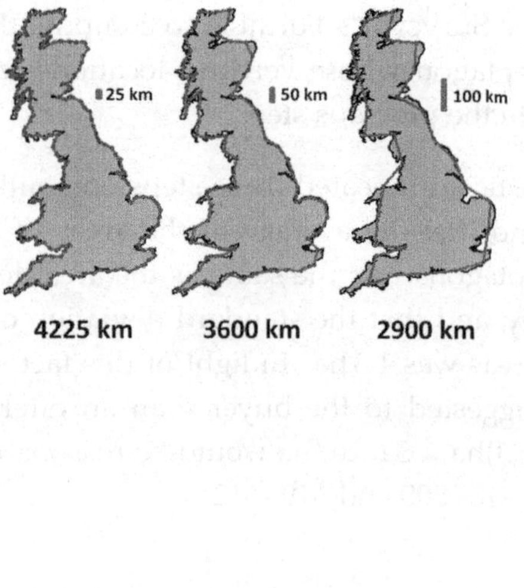

The estimated length of the UK coastline depends on the size of the ruler used (modified from Gurung [2017]).

Weighing

A laboratory weight C, of nominal mass 200 g, is to be calibrated using two previously calibrated reference weights A and B, whose masses are 200.000 22 g and 200.000 61 g, respectively, both within 0.000 14 g of their true masses with 95 % probability. This suggests that these may be class E_2 weights.[39]

Radwag AK-4/2000 Automatic Mass Comparator (Radom, Poland).

The calibration involves determining three mass differences using a mass comparator: the observed difference between the masses of A and B is $D_{AB} = -0.38$ mg, and similarly $D_{AC} = -1.59$ mg and $D_{BC} = -1.22$ mg.

Since the weight A has a nominal mass 200 g, we write $m_A = 200\,\text{g} + \delta_A$, where δ_A is the true deviation from the nominal mass. Using the same notation for the other weights, we have the following simultaneous *observation equations*:[40]

$$D_{AB} = \delta_A - \delta_B + \varepsilon_{AB},$$
$$D_{AC} = \delta_A - \delta_C + \varepsilon_{AC},$$
$$D_{BC} = \delta_B - \delta_C + \varepsilon_{BC},$$

where ε_{AB}, ε_{AC}, and ε_{BC} denote the (non-observable) measurement errors incurred in the mass comparator. The conventional approach[41] involves finding values for δ_A, δ_B, and δ_C, that minimize the sum of the squared errors,

$$(\delta_A - \delta_B - D_{AB})^2 + (\delta_A - \delta_C - D_{AC})^2 + (\delta_B - \delta_C - D_{BC})^2,$$

subject to the constraint $\delta_A + \delta_B = 0.83$ mg, which is one of several alternative constraints that could be applied.

The solution of this constrained linear least squares (Page 196) problem produces the estimate $\widehat{\delta}_C = 1.82$ mg, with associated uncertainty $u(\widehat{\delta}_C) = 0.05$ mg. Even though the maximum permissible error for a 200 mg class E_1 weight is 0.10 mg, it would be inappropriate to place the weight C into this class, considering that the

[39] International Organization of Legal Metrology (OIML). *Weights of classes E_1, E_1, E_2, F_1, F_2, M_{1-2}, M_2, M_{2-3}, and M_3 — Part 1: Metrological and technical requirements.* Bureau International de Métrologie Légale (OIML), Paris, France, 2004. URL https://www.oiml.org/en/files/pdf_r/r111-1-e04.pdf. International Recommendation OIML R 111-1 Edition 2004 (E)

[40] P. E. Pontius and J. M. Cameron. *Realistic Uncertainties and the Mass Measurement Process — An Illustrated Review.* Number 103 in NBS Monograph Series. National Bureau of Standards, Washington, DC, 1967. URL http://nvlpubs.nist.gov/nistpubs/Legacy/MONO/nbsmonograph103.pdf

[41] R. N. Varner and R. C. Raybold. *National Bureau of Standards Mass Calibration Computer Software.* NIST Technical Note 1127. National Bureau of Standards, Washington, DC, July 1980. URL https://nvlpubs.nist.gov/nistpubs/Legacy/TN/nbstechnicalnote1127.pdf

calibrants are class E_2 weights.

Alternatively, an estimate of δ_C can be obtained using Bayesian statistical methods. For this, we model the measured mass differences probabilistically, as outcomes of Gaussian random variables (Page 167):

$$D_{AB} \sim \text{GAU}(\delta_A - \delta_B, \sigma),$$
$$D_{AC} \sim \text{GAU}(\delta_A - \delta_C, \sigma),$$
$$D_{BC} \sim \text{GAU}(\delta_B - \delta_C, \sigma).$$

For example, the observed value of D_{AB} is viewed as a drawing from a Gaussian distribution with mean $\delta_A - \delta_B$ and standard deviation σ. We also use probability distributions to express the uncertainty about these deviations from the nominal masses of weights A and B, thus:

$$\delta_A \sim \text{GAU}(0.22\text{ mg}, 0.07\text{ mg}),$$
$$\delta_B \sim \text{GAU}(0.61\text{ mg}, 0.07\text{ mg}).$$

All we know about weight C is that it has a nominal mass of 200 g, but we also have good reasons to believe that its true mass lies within a reasonably narrow interval centered at 200 g. Providing a generous allowance for the length of this interval, we adopt the model

$$\delta_C \sim \text{GAU}(0\text{ mg}, 100\text{ mg}).$$

The fact that this prior standard deviation, 100 mg, is comparable to the maximum permissible error for a class M_3 weight, does not signify that the weight C may be of this class. Rather, this choice serves only to give the data ample opportunity to make themselves heard, unencumbered by overly restrictive prior assumptions.

Since the Bayesian approach (Page 204) requires that all unknown parameters be modeled probabilistically, we need to assign a probability distribution also to the standard deviation, σ, of the measurement errors. Here we assume that the true value of σ is *a priori* equally likely to be larger or smaller than 1 mg, and assign a half-Cauchy distribution (Page 170) to σ, with median 1 mg. This choice provides considerable latitude for the value that σ may truly have.

The following implementation of the Bayesian model in the Stan language will be assumed to have been assigned to `model` as a character string (including line breaks), before executing the subsequent R code.

The Monte Carlo Markov Chain method (Page 209), implemented using the Stan modeling language in tandem with the R package `rstan` as detailed alongside, was used to draw a large sample from the posterior probability distribution of δ_C. A robust estimate of the mean of this sample equals 1.82 mg (which happens to be identical to the least squares estimate above), and a robust estimate of its standard deviation equals 0.07 mg, which is appreciably larger than the uncertainty associated with the least squares estimate.

```
model = "
 data {
   real DAB;
   real DAC;
   real DBC;
 }
 parameters {
   real dA;
   real dB;
   real dC;
   real<lower=0> sigma;
 }
 model {
   // Prior distributions
   dA ~ normal(0.22, 0.07);
   dB ~ normal(0.61, 0.07);
   dC ~ normal(0.00, 10);
   sigma ~ cauchy(0.0, 1);
   // Likelihood
   DAB ~ normal(dA - dB, sigma);
   DAC ~ normal(dA - dC, sigma);
   DBC ~ normal(dB - dC, sigma);
 }
 "
```

```
require(rstan)
require(robustbase)
fit = stan(model_code = model,
        data = list(DAB = -0.38, DAC = -1.59,  DBC = -1.22),
        warmup=75000, iter=250000,
        chains=4, cores=4, thin=25,
        control= list(adapt_delta=0.999))

dC.posterior = rstan::extract(fit)$dC
c(MEAN=huberM(dC.posterior)$mu, SD=Qn(dC.posterior))
```

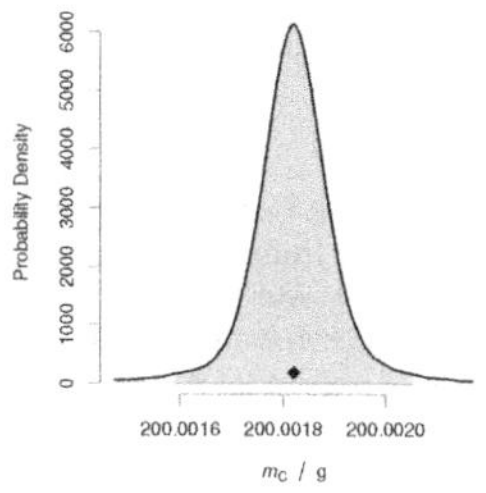

Probability density (Page 159) for m_C produced by the Bayesian calibration. Its mean value (indicated by a diamond), is the calibrated value of m_C. The horizontal, dark, thick line segment indicates an interval of half-length 0.23 mg that, with 95 % probability, is believed to include the true value of m_C.

The results of this calibration are $m_C = 200\,001.82$ mg, give or take 0.23 mg, with 95 % probability.

Optimal design of experiments can use the results of uncertainty propagation as a guide. Consider a situation where we wish to determine the individual weights of three gold coins with the smallest uncertainty possible. We have access to a good balance but only for a limited time, enough to perform three weighings.

We assume that the uncertainty associated with each weighing in this balance is constant and does not depend on the mass being weighed, $u(m) = u$, for values of mass within the range under consideration.

We could devise two experimental designs: (i) weigh each coin individually or (ii) weigh them in pairs (coin 1 and coin 2 together, then coin 1 and coin 3 together, and finally coins 2 and 3 together). This is the measurement model corresponding to the latter design:

$$m_1 = \tfrac{1}{2}\big(+ m_{1+3} + m_{1+2} - m_{2+3}\big),$$
$$m_2 = \tfrac{1}{2}\big(- m_{1+3} + m_{1+2} + m_{2+3}\big),$$
$$m_3 = \tfrac{1}{2}\big(+ m_{1+3} - m_{1+2} + m_{2+3}\big).$$

Applying Gauss's formula to these expressions yields, for example,

Since the expressions above are linear combinations of the weighings, Gauss's formula is exact in this case.

$$\begin{aligned} u^2(m_1) &= \left(\frac{\partial m_1}{\partial m_{1+3}}\right)^2 u^2(m_{1+3}) + \left(\frac{\partial m_1}{\partial m_{1+2}}\right)^2 u^2(m_{1+2}) \\ &\quad + \left(\frac{\partial m_1}{\partial m_{2+3}}\right)^2 u^2(m_{2+3}) \\ &= \tfrac{1}{4}u^2 + \tfrac{1}{4}u^2 + \tfrac{1}{4}u^2, \end{aligned}$$

and similarly for $u(m_2)$ and $u(m_3)$. Thus,

$$u(m_1) = u(m_2) = u(m_3) = u\sqrt{3/4}.$$

Hence, by weighing the three coins in pairs we can achieve 13 % lower uncertainty than by weighing them separately.

Similarly to how the previous example showed that clever experimental designs can lead to better measurements, the next example illustrates how, by taking uncertainty into account, one can build a better resistor.

Consider four resistors, each with nominal resistance of 100 kΩ give or take 1 kΩ, arranged in a circuit as shown alongside. If the connections have negligible resistance, then the resistance of such circuit is

$$R = \frac{1}{\frac{1}{R_1+R_2} + \frac{1}{R_3+R_4}} = 100\,\mathrm{k\Omega}.$$

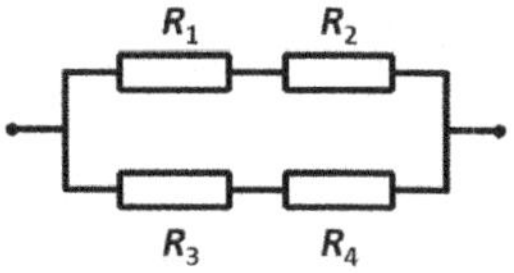

A circuit comprising two pairs of resistors in parallel, with each pair connected in series, all with resistance 100 ± 1 kΩ, will also have 100 kΩ resistance but with uncertainty that is two times smaller than the uncertainty of each individual resistor.

Suppose we model the resistances as independent random variables that are distributed uniformly (Page 167) between 99 kΩ and 101 kΩ, implying that the standard uncertainty of each resistor in the circuit is $(101\,\mathrm{k\Omega} - 99\,\mathrm{k\Omega})/\sqrt{12} = 0.6\,\mathrm{k\Omega}$.

While the resistance, $R = 100\,\mathrm{k\Omega}$, of the circuit remains the same as the resistance of the individual resistors, propagating their uncertainties, for example using the *NIST Uncertainty Machine*, shows that the standard uncertainty associated with R, $u(R) = 0.3\,\mathrm{k\Omega}$, is half the standard uncertainty of the individual resistors, and has a probability distribution that is close to a Gaussian distribution.

Although many factors can contribute to the uncertainty of a resistor, including changes in their temperature, ambient humidity, or even their age, it is possible to build a circuit of nominally identical resistors that has the same resistance as the individual resistors, but is better than any of them, in the sense that it dissipates power more efficiently than a single resistor of the same resistance would.

Ranking

Ranking is assigning a place for an object being measured in an ordered sequence of standards, based on the value of a property whose values can be ordered from smallest to largest but not necessarily quantified. For example, to distinguish harder and softer pencil leads, pencil manufacturers rank pencils on a grading scale: from 9B (super black, very soft) to 9H (a gray scratch, very hard).

HARDNESS	MINERAL
1	talc
2	gypsum
3	calcite
4	fluorite
5	apatite
6	orthoclase
7	quartz
8	topaz
9	corundum
10	diamond

The minerals defining the Mohs hardness scale.

THE MOHS HARDNESS SCALE is determined by comparing a mineral specimen against a set of reference standards by means of a scratch test, whose results place it in the rank order of increasing hardness. The Mohs reference standards [Klein and Dutrow, 2007] are samples of various minerals with ordinal values 1 to 10 assigned to them without implying that the increase in hardness from gypsum to calcite is the same as the increase in hardness from apatite to orthoclase. For example, tourmaline typically scratches quartz and is scratched by topaz, hence its Mohs hardness is between 7 and 8. The numbers used to denote ranking order on an ordinal scale are nothing but labels for which arithmetic operations are not meaningful. Thus, numbers 1–10 could very well be replaced by letters A–J to convey the same message. In practice, when one says that the hardness of tourmaline is 7.5, all one means is that its hardness lies between the hardness of quartz and topaz.

Numbers are often used as labels with only an ordinal or nominal connotation. Examples of this use are the numbers used in the Saffir-Simpson ordinal scale of hurricane strength, and the numbers printed on the shirts of football players, where they have a nominal role.

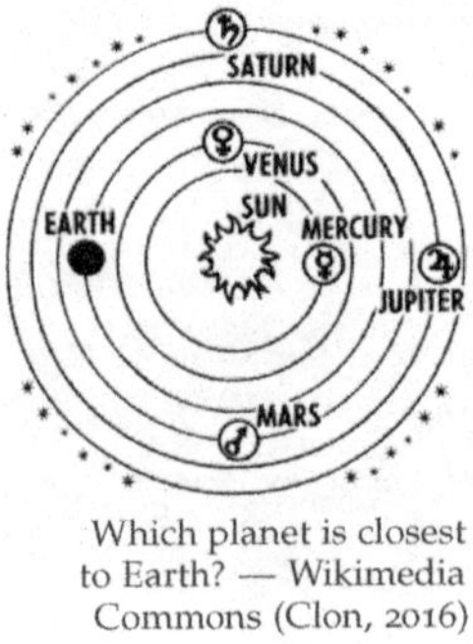

Which planet is closest to Earth? — Wikimedia Commons (Clon, 2016)

OUR ANCESTORS HAVE PONDERED for ages the question of which planet is the closest to Earth. Most textbooks state that it is Venus because it makes the closest approach to Earth compared to any other planet [Stockman et al., 2019]. The answer, however, depends on what is meant by "closest" — whether it means closest ever, closest on average, or closest most of the time —, because planets do not stand still and therefore distances between them are in constant flux.

On January 1st, 2019, for example, Venus indeed was the planet closest to Earth, but that was no longer the case two months later when Mercury moved closer.

In the long term (over the period 2020–2420) Mercury will be Earth's closest neighbor 47 % of the time, Venus 37 % of the time, and Mars 16 % of the time, according to the NASA Jet Propulsion Laboratory HORIZONS system [Giorgini, 2015]. It may even be surprising that Pluto will be closer to Earth than Neptune 4 % of the time, even though its median distance to Earth is almost 1.5 times larger than Neptune's.

To characterize the positions of the planets relative to Earth properly, one needs to consider the distributions of the daily distances, as depicted in the histograms below.

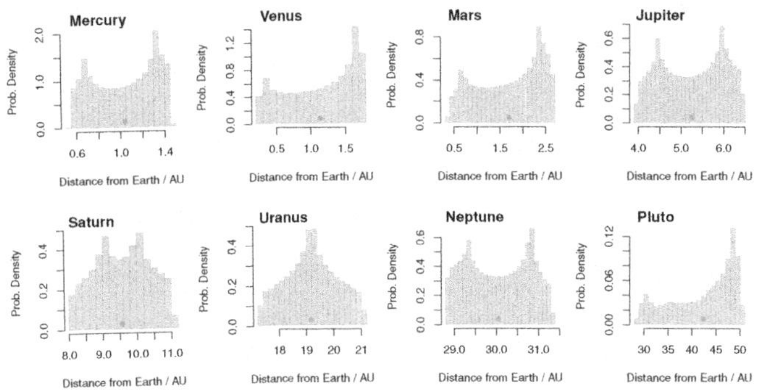

Histograms of the daily distances from Earth (expressed in astronomical unit, AU, which is the mean distance between Sun and Earth, approximately 150×10^6 km), for the planets in the Solar System during the period 2020–2420. Each dot indicates the average distance from Earth.

Except for Uranus, the average distance does not represent a typical distance from Earth. Neither does the standard deviation of the daily distances capture the variability of the distances accurately.

Even though the uncertainty of the distance from Earth to any other planet, computed for a particular day by the HORIZONS system, is rather small, the variability of the distances over time is quite large, and it is best communicated by means of the probability distributions depicted in these histograms, which may be interpreted as representing the uncertainty about the distance on a randomly selected day.

In the 2022 Winter Olympics, the gold and silver medals in pairs figure skating were awarded to Wenjing Sui & Cong Han and Evgenia Tarasova & Vladimir Morozov, respectively, who earned total scores of 239.88, and 239.25 points, from a panel of nine judges.

CBC Sports
@cbcsports
Sui Wenjing and Han Cong edged their Russian opponents by 0.63 points to claim gold on home ice at #Beijing2022

CBC Sports tweet on Feb 18, 2022.

The medals are awarded considering only the final ranking of the athletes, regardless of whether the differences in the underlying scores are large or small. In 2022, a mere 0.63 point gap (that is, 0.26 %) separated Olympic gold from silver. How significant may this difference be considering the uncertainty that inevitably is associated with the assignment of scores?

In the 2018 Winter Olympics, the difference between the gold and silver medals in the ladies single skating was 1.31 points with the 15-year-old Alina Zagitova from Russia claiming the gold.

	EXECUTED ELEMENT		
JUDGE	5RLi4	3Li4	3S
1	4	4	1
2	3	4	0
3	3	5	1
4	3	4	1
5	1	2	1
6	2	3	1
7	3	4	2
8	3	4	1
9	3	4	2
Base value, b	7.00	5.10	4.30
Weight, w	0.70	0.51	0.43
Total, s	9.00	7.07	4.79

Excerpt of the score sheet for the final free skating component by Tarasova and Morozov at the 2022 Olympics. 3S stands for triple Salchow whereas 3Li4 and 5RLi4 for various lifts of certain difficulty.

Figure skating scores are produced by a complex scoring system that involves intrinsic levels of difficulty for technical elements, a priori weights, subjective evaluations made by nine judges independently of one another, and consideration of whether the elements are performed early or late during each routine.

The Monte Carlo method — that is, a method based on simulations of contributions from recognized sources of uncertainty — can be used to carry out uncertainty evaluations, and, in this case, it will also shed light on the significance of the difference in scores that separated the silver and gold medals.

The previous table shows an excerpt from the score

sheet for the free skating component in the Olympic finals: each executed technical element, i, has a particular, agreed-upon base value, b_i, and the quality of its execution is evaluated by nine judges. After removing the lowest and highest scores, the average score of the other seven is computed (trimmed mean) and added to the base value after multiplication by a predetermined weight, w_i. The combined score for element i is computed as follows, where $J_{i,j}$ denotes the score that judge j gave the athlete for the execution of this element:

$$s_i = b_i + \frac{w_i}{9-2}\left(\sum_{j=1}^{9} J_{i,j} - \min_{j=1,\ldots,9}\{J_{i,j}\} - \max_{j=1,\ldots,9}\{J_{i,j}\}\right).$$

The final scores are the sums of such element-specific scores, and certainly include expressions of the subjective, professional opinions of the nine judges. Given that judges do not always agree on their scores, it is reasonable to explore the extent of their disagreement.

One way to assess the reliability of the judging scores is to simulate samples by randomly drawing scores, with replacement, from the set of actually observed scores, and then calculating the total score for each such random sample. This method is known as the nonparametric bootstrap [Efron and Tibshirani, 1993]: it is widely used for uncertainty evaluations in science, medicine, and engineering. In this case, we generated 100 000 bootstrap samples, which enabled us to conclude that the probability of Tarasova & Morozov winning the gold medal was 23 %, thus quantifying the effect that judging uncertainty had upon the final result.

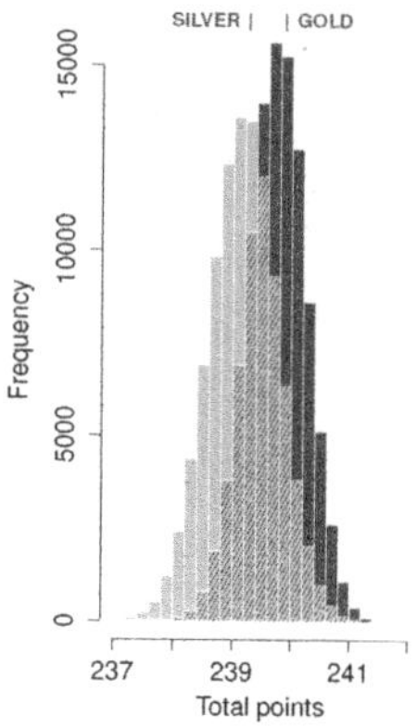

Probabilistic interpretation of the pairs figure skating gold and silver medal scores at the 2022 Winter Olympics.

Comparing

One of the most important applications of uncertainty evaluation is to compare two quantities whose measured values are surrounded by uncertainty.

There is no margin for doubt when comparing numbers about which there is no uncertainty: everyone agrees that $9 > 7$. But it is impossible to decide conclusively whether the meitnerium-277 and meitnerium-278 isotopes have the same or different longevity, considering that their half-lives are estimated as $t_{½}(^{277}\mathrm{Mt}) = 9\,\mathrm{s}$ and $t_{½}(^{278}\mathrm{Mt}) = 7\,\mathrm{s}$ with standard uncertainties 6 s and 3 s, respectively [Audi et al., 2017].

We shall illustrate six kinds of comparisons:

(1) a single value of a property of a reference material measured by a user of the material, against the corresponding certified value;

(2) several replicated determinations of the value of a property of a reference material against the certified value of the same property;

(3) a set of replicated observations of the value of a quantity against a specified target value that the quantity is supposed to have;

(4) uncertainty components expressing variability between and within the individual units of a reference material.

(5) a time series of observations against a threshold;

(6) two methods for measuring the same property.

Comparing Measured Value with Reference Value

When comparing a measured value and a certified value, while taking their uncertainties into account, the overlap of corresponding coverage intervals is not sufficient reason to conclude that the corresponding true values are identical [Possolo, 2020, Example 7.2.A].

Note that equality to within specified uncertainties is not a transitive relation. Thus, if objects A and B are found to have identical masses to within their uncertainties, and if the same is true for objects B and C, it does not necessarily follow that the masses of A and C also are identical to within their respective uncertainties.

The certified mass fraction of nickel in NIST Standard Reference Material (SRM) 59a (ferrosilicon) is 328 mg/kg with expanded uncertainty 73 mg/kg for 95 % coverage. The Bayesian interpretation (Page 205) of this fact is that the corresponding true value is believed to lie between 255 mg/kg and 401 mg/kg with 95 % probability.

Suppose that a user of this material has measured the mass fraction of nickel and obtained 172 mg/kg with expanded uncertainty 132 mg/kg, also for 95 % coverage. Since the corresponding coverage interval, ranging from 40 mg/kg to 304 mg/kg, overlaps the interval above, the inference could be drawn that there is no significant difference between the true mean of the user's measurement and the true value of the measurand.

The difference between the two measured values is 328 mg/kg − 172 mg/kg = 156 mg/kg, and the standard uncertainty of the difference between these values is the square root of the sum of the individual, squared standard uncertainties,

$$\sqrt{\left(\tfrac{1}{2}\times 73\ \mathrm{mg/kg}\right)^2 + \left(\tfrac{1}{2}\times 132\ \mathrm{mg/kg}\right)^2} = 75\ \mathrm{mg/kg}.$$

This statistical test assumes that the two values being compared are outcomes of independent Gaussian random variables, and that their associated standard uncertainties are based on vary large numbers of degrees of freedom.

The p-value (Page 32) is the probability of observing a difference as large or larger (in absolute value) than the difference that was observed, by chance alone, owing to the vagaries of sampling and measuring the material, if the corresponding true values were identical. A small p-value suggests a significant difference.

The test statistic for whether this difference is significantly different from zero is the standardized difference, (156 mg/kg)/(75 mg/kg) = 2.08. The p-value of this test is the probability of a Gaussian random variable with mean 0 and standard deviation 1 being either smaller than −2.08 or larger than +2.08. This probability is 3.75 %, which is usually interpreted as suggesting a significant difference.

Comparing Replicated Determinations with Reference Value

To validate a measurement method, a laboratory often makes measurements of a reference material, and then compares the measurement results with the certified value. The NIST SRM 1944 is a mixture of marine sediments collected near urban areas in New York and New Jersey, intended for use in evaluations of analytical methods for the determination of polychlorinated biphenyls (PCBs) and other hydrocarbons in similar matrices.

The hypothesis of no difference between measured and certified values entails that the criterion t should be like an outcome from a Student's t distribution with 5.3 degrees of freedom. The larger the absolute value of t is, the more surprising it is that it should have occurred by chance alone, without there actually being a difference between measured and certified values. The questionable "logic" behind conventional tests of hypotheses is that rare events should not happen. Here, however, the probability is 15 % that an absolute value of 1.7 or larger might happen by chance alone owing to the vagaries of sampling, a far cry from a rare event, hence the conclusion that there is insufficient reason to reject the hypothesis of equality between measured and certified values.

A quality control test yielded the following replicates for the mass fraction of PCB 95:

$$63.9\,\mu\text{g/kg}, \quad 48.4\,\mu\text{g/kg}, \quad \text{and} \quad 46.1\,\mu\text{g/kg}.$$

Their average and standard deviation are 52.8 µg/kg and 9.7 µg/kg. The Type A evaluation of the standard uncertainty associated with the average is $(9.7\,\mu\text{g/kg})/\sqrt{3} = 5.6\,\mu\text{g/kg}$, on 2 degrees of freedom.

The certified mass fraction of PCB 95 in SRM 1944 is 65.0 µg/kg, with standard uncertainty 4.45 µg/kg. The comparison criterion is

$$t = \frac{52.8 - 65.0}{\sqrt{5.6^2 + 4.45^2}} = -1.7.$$

On the hypothesis of no difference between the mean of the laboratory results and the certified value, this should be approximately like an outcome of a Student's t random variable with effective number of degrees of freedom (ν) given by the Welch-Satterthwaite formula [JCGM 100:2008, G.4]:

$$\nu = \frac{(5.6^2 + 4.45^2)^2}{\dfrac{5.6^4}{2} + \dfrac{4.45^4}{7}} = 4.8.$$

The "7" appearing in the denominator is the "effective" number of degrees of freedom associated with the uncertainty evaluation for the certified value because the corresponding certificate suggests that 8 measurement results was used to determine this value.

Since the probability is 15 % that such random variable will deviate from 0 by more than 1.7 standard deviations, we conclude that the laboratory measurements do not differ significantly from the certified value.

This conclusion is contingent on the three replicated determinations the laboratory made being like a sample from a Gaussian distribution — an assumption that is next to impossible to verify reliably with so few observations. Still, the Shapiro-Wilk test of Gaussian shape, whose R implementation accommodates samples this small, yields a comforting *p*-value (Page 32) of 23 %.

Comparing Replicated Determinations with Target Value

A particular kind of artillery shell is supposed to be loaded with 333 g of propellant. The values of the mass of propellant in 20 such shells, expressed in gram, were:

$$295,\ 332,\ 336,\ 298,\ 300,\ 337,\ 307,\ 312,\ 301,\ 333,$$
$$344,\ 340,\ 339,\ 341,\ 297,\ 335,\ 345,\ 342,\ 322,\ 331.$$

The conventional treatment of this problem[42] involves computing the difference between the average of these 20 determinations, 324 g, and the specified target value, and expressing it relative to the standard uncertainty of the average:

[42] M. G. Natrella. *Experimental Statistics*. National Bureau of Standards, Washington, D.C., 1963. National Bureau of Standards Handbook 91

$$t = \frac{324\,\mathrm{g} - 333\,\mathrm{g}}{18.3\,\mathrm{g}/\sqrt{20}} = -2.2.$$

The denominator has the standard deviation of the determinations, 18.3 g, divided by the square root of the number of determinations, which is the Type A evaluation of standard uncertainty for the average, according to the GUM (4.2.3). Therefore, the average of these determinations is 2.2 standard uncertainties below the specified target value.

Still according to the conventional treatment, this standardized difference is to be interpreted as if it were an outcome of a Student's t distribution with 19 degrees of freedom. The probability that such random variable will take a value that is more than 2.2 units away from zero, in either direction, is 4 %.

The reason why we consider deviations from zero in either direction is that we are testing a difference between the mean of the measured values and the specified value, regardless of whether that mean is larger or smaller than this specified value.

The p-value (Page 32) of a two-sided Student's t test can be calculated using a variety of software. Since any software may suffer from errors, it is recommended that important calculations be replicated using implementations developed independently of one another in different software environments.

That probability, 4 %, is called the p-value of the test (Page 32). It is the probability of observing a difference at least as large, in absolute value, as the difference that was observed, owing to the vagaries of sampling alone, on the assumption that in fact there is no difference. For this reason, a small p-value is usually interpreted as suggesting that the observed difference is significant.

The test just described is a procedure for statistical inference: the derivation of a conclusion from a sample, where the confidence in the conclusion is characterized probabilistically. The validity of the results of all such procedures hinges on the adequacy of the model and on particular assumptions, which are much too often neglected or taken for granted.

In this case, the assumptions are that the values in the sample are like outcomes of independent, Gaussian random variables, all with the same mean and standard deviation.

Whereas independence (Page 163) is a powerful property, it is also next to impossible to verify empirically in most cases. As for the Gaussian shape, the Anderson-Darling test yields a p-value of 0.002 (Page 32), indicating that the data are rather unlikely to be a sample from a Gaussian distribution.[43]

[43] T. W. Anderson and D. A. Darling. Asymptotic theory of certain "goodness-of-fit" criteria based on stochastic processes. *Annals of Mathematical Statistics*, 23:193–212, 1952. doi:10.1214/aoms/1177729437

```
m = c(295, 297, 298, 300, 301, 307, 312, 322, 331, 332,
      333, 335, 336, 337, 339, 340, 341, 342, 344, 345)
library(nortest)
ad.test(m)$p.value
```

This suggests that the Student's t test may not be appropriate for these data, and that conformity with the target value ought best be evaluated in some other way.

Unlike the Student's t test, the Wilcoxon's one-sample signed rank test[44] does not require that the replicate determinations be like a sample from a Gaussian distribution, only that the distribution be symmetric. The corresponding p-value is 0.22 (Page 32):

[44] M. Hollander, D. A. Wolfe, and E. Chicken. *Nonparametric Statistical Methods*. John Wiley & Sons, Hoboken, NJ, 3rd edition, 2014. ISBN 978-0-470-38737-5

```
wilcox.test(m, mu=333)$p.value
```

Therefore, the result of this test contradicts the result of Student's t test above and suggests that the observations are consistent with the target value of 333 g.

This example shows that conclusions drawn from data depend on assumptions and models used to describe particular patterns of variability of the data, and that the conclusions may change drastically when assumptions or models change.

In 2014, 29 teams of researchers were asked to analyze the same data about red cards in soccer, using statistical procedures of their choice. Twenty teams concluded that there is a significant correlation between a player's skin color and his being given a red card, whereas nine teams concluded that there is none [Silberzahn and Uhlmann, 2015].

Comparing Uncertainty Components for a Reference Material

Assessing the homogeneity of a candidate reference material involves comparing the variability of the values of a property between units of the material, with their variability within units.

The NIST SRM 2684c is a bituminous coal intended primarily for evaluations of analytical methods used for coals. Each unit of the material is a bottle containing 50 g of the finely powdered material.

Between two and four aliquots from each of 23 selected bottles of the material were analyzed by X-ray fluorescence spectroscopy for aluminium content.

The conventional assessment of homogeneity is based on a statistical technique called *analysis of variance* (ANOVA).[45]

[45] R. A. Fisher. *Statistical Methods for Research Workers*. Hafner Publishing Company, New York, NY, 14th edition, 1973

Since the measurement results appear to be consistent with the modeling assumptions that validate ANOVA, it may be worth pointing out that the results from this technique suggest that the material is significantly heterogeneous: the F-test yields p-value 0.045 (Page 32).

```
z = data.frame(
 bottle=c("B01", "B01", "B02", "B02", "B03", "B03", "B04", "B04",
          "B05", "B05", "B06", "B06", "B07", "B07", "B08", "B08",
          "B09", "B09", "B10", "B10", "B11", "B11", "B12", "B12",
          "B12", "B12", "B13", "B13", "B14", "B14", "B15", "B15",
          "B16", "B16", "B17", "B17", "B18", "B18", "B19", "B19",
          "B19", "B20", "B20", "B21", "B21", "B22", "B22", "B23",
          "B23"),
 kcps=c(62.37, 61.18, 60.73, 60.03, 60.91, 60.59, 60.71, 61.15,
        60.39, 60.59, 61.30, 61.32, 61.09, 61.14, 60.50, 61.91,
        60.47, 60.15, 61.39, 61.07, 61.08, 60.34, 60.51, 60.66,
        61.81, 61.09, 61.79, 60.97, 60.96, 61.23, 60.33, 59.90,
        59.91, 60.30, 61.30, 60.81, 60.83, 61.00, 60.30, 60.49,
        60.55, 62.24, 60.90, 60.61, 60.80, 60.69, 60.91, 60.78,
        60.94))

z.aov = aov(kcps~bottle, data=z)
summary(z.aov)
qqnorm(residuals(z.aov))
```

Next we will employ a model-based approach to evaluate potential heterogeneity, which is quantified by a parameter in the measurement model. The model, which will reappear in the discussion of *Consensus Building* (Page 146), expresses the fluorescence intensity attributable to aluminium as

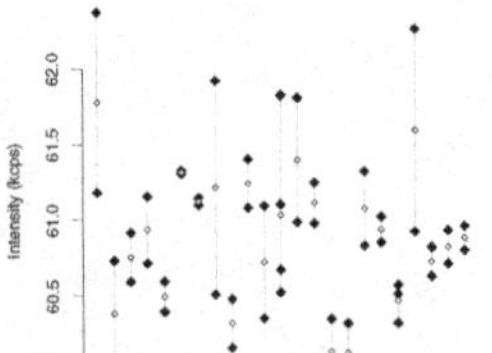

X-ray fluorescence intensity from Al in aliquots drawn from bottles of NIST SRM 2684. Each open diamond represents the average of the determinations made in aliquots from the same bottle.

$$I_{ij} = \mu + \beta_j + \varepsilon_{ij},$$

where $j = 1, \ldots, n$ ($n = 23$) denotes the bottle number, $i = 1, \ldots, m_j$ denotes the aliquot (subsample) drawn from bottle j, μ is the overall mean intensity, β_j denotes the effect of bottle j, and ε_{ij} denotes the effect of aliquot i from bottle j. Only the $\{I_{ij}\}$ are observable.

The bottle effects, $\{\beta_j\}$, are modeled as outcomes of random variables all with mean zero and standard deviation τ, and the aliquot effects $\{\varepsilon_{ij}\}$ are modeled as outcomes of random variables all with mean zero and standard deviation σ. These random variables do not need to be independent: it suffices that the bottle effects among themselves, and the aliquot effects among themselves, be *exchangeable* (Page 164).

Material whose aluminium content is homogeneous should exhibit no significant differences between bottles above and beyond the differences between aliquots from the same bottle, and τ will not differ significantly from zero: this means that readings of fluorescence intensity in aliquots from different bottles are *not* more variable than readings in aliquots from the same bottle.

Suppose that $t = T(\mathbf{I})$ is an estimate of τ, where $\mathbf{I}$ denotes the set of 49 observations of fluorescence intensity, together with a description of which aliquots go with which bottles. The function T computes an estimate of τ taking into account the structure of the data. Small values of t suggest that the material is homogeneous, and large values suggest that it is not.

Permute the elements of $\mathbf{I}$ randomly, similarly to how one would shuffle a deck of playing cards, so that the value of a particular aliquot from a particular bottle may take the place of the value of any other aliquot, from any other bottle, the result being $\mathbf{I}^*$. If the material really is homogeneous, then $t^* = T(\mathbf{I}^*)$ should be close to t.

Now, imagine repeating this process a large number K of times, thus obtaining $t_1^*, \ldots, t_K^*$. The dispersion of these values reflects the variability of estimates of τ that is to be expected owing to the vagaries of sampling alone, on the assumption that the material indeed is homogeneous.

Finally, compare the value of t that corresponds to the actual data, with the set $\{t_k^*\}$, and determine how "unusual" t may be among the $\{t_k^*\}$. If t should be unusually large, then this may warrant concluding that the material is heterogeneous.

There are many different ways of estimating τ, and it does not matter very much which one we will choose. For this example, we will rely on one of the most widely used estimators of τ — the restricted maximum likelihood estimator (REML).[46]

[46] S. R. Searle, G. Casella, and C. E. McCulloch. *Variance Components*. John Wiley & Sons, Hoboken, NJ, 2006. ISBN 0-470-00959-4

We will compute the value of τ that corresponds to the actual measurement data (what above we called t), and we will also compute the values of τ for each of $K = 10\,000$ permutations of the data (what above we called $\{t_k^*\}$). The R code listed below implements this *permutation test*.[47]

[47] K. J. Berry, J. E. Johnston, and Jr. P. W. Mielke. *A Primer of Permutation Statistical Methods*. Springer, Cham, Switzerland, 2019. ISBN 978-3-030-20932-2. doi:10.1007/978-3-030-20933-9

Out of 9990 permutations of the data (for 10 permutations the estimation procedure did not converge), only 458 yielded an estimate of τ that is larger than the estimate obtained for the actual data ($\tau = 0.31$ kcps).

Therefore, the p-value (Page 32) of the permutation test of homogeneity is $458/9990 = 4.6$ %, which is commonly regarded as suggesting that the material is not homogeneous, in a conventional statistical test of the hypothesis of homogeneity whose probability of erroneously rejecting this hypothesis is set at 5 %.

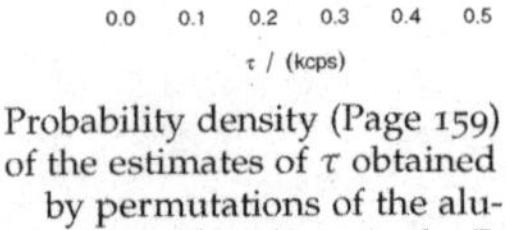

Probability density (Page 159) of the estimates of τ obtained by permutations of the aluminium data (tauB in the R code alongside): the diamond indicates the estimate of τ for the original data. The shaded area amounts to 4.6 % of the area under the curve.

```
library(nlme)
z.lme = lme(kcps~1, random=~1|bottle, data=z, method="REML")
summary(z.lme)
intervals(z.lme)

tau = as.numeric(VarCorr(z.lme)["(Intercept)", "StdDev"])
K = 10000; zB = z; tauB = rep(NA, K);
for (k in 1:K)
{ zB$kcps = sample(z$kcps, size = nrow(z), replace = FALSE)
  zB.lme = try(lme(kcps ~ 1, random = ~1|bottle,
                        data = zB, method = "REML"))
  if (class(zB.lme) == "try-error") { next }
  else { tauB[k] = as.numeric(
    VarCorr(zB.lme)["(Intercept)", "StdDev"]) }
}
tauB = tauB[complete.cases(tauB)]
## p-value
sum(tauB > tau) / length(tauB)
```

If one is willing to make additional assumptions about the bottle effects and about the aliquot effects, for example that they are like outcomes of independent, Gaussian random variables with standard deviation τ for the $\{\beta_j\}$ and σ for the $\{\varepsilon_{ij}\}$, then `summary(z.lme)` and `intervals(z.lme)` will produce not only the aforementioned estimate of τ, but also an approximate 95 % coverage interval for its true value ranging from 0.16 kcps to 0.60 kcps, which suggests heterogeneity.

The question of homogeneity can also be answered using other statistical procedures that do not make particular distributional assumptions, the same as the permutation test considered above. For example, the Kruskal-Wallis test,[48] carried out as

48 M. Hollander, D. A. Wolfe, and E. Chicken. *Nonparametric Statistical Methods*. John Wiley & Sons, Hoboken, NJ, 3rd edition, 2014. ISBN 978-0-470-38737-5

```
kruskal.test(kcps~bottle, data=z)
```

yields p-value 0.0522 (Page 32).

The same statistical model, $I_{ij} = \mu + \beta_j + \varepsilon_{ij}$, can also be fit to the data using a Bayesian (Page 204) procedure, which involves distributional assumptions about the $\{\beta_j\}$ and about the $\{\varepsilon_{ij}\}$, and requires the specification of prior distributions.

The R package `brms`[49] provides a user friendly way to implement a wide variety of Bayesian models, including the model we are concerned with here.

49 P.-C. Bürkner. Advanced Bayesian multilevel modeling with the R package brms. *The R Journal*, 10(1):395–411, 2018. doi:10.32614/RJ-2018-017

```
z.brm = brm(kcps ~ 1 + 1|bottle, data=z,
            iter=5e5, warmup=1e5, thin=25, cores=4))

z.brm.mcmc = as.mcmc(z.brm, combine_chains=TRUE)
tau = z.brm.mcmc[, "sd_bottle__Intercept"]
quantile(tau, probs=c(0.025, 0.5, 0.975))
```

The prior distributions that function `brm` uses by default for this model are all Student's t_3, with those pertaining to τ and to σ truncated at 0.

The median of the posterior distribution of τ produced by this approach is $\tau = 0.29$ kcps, and a 95 % credible interval for its true value ranges from 0.03 kcps to 0.53 kcps, again suggesting heterogeneity.

Comparing a Time Series against a Threshold

> On May 9, 2013, CO_2 levels in the air reached the level of 400 parts per million (ppm). This is the first time in human history that this milestone has been passed [. . .] To some, crossing the threshold of 400 ppm is a signal that we are now firmly seated in the 'Anthropocene,' a human epoch where people are having major and lasting impacts on the planet. — NASA Global Climate Change

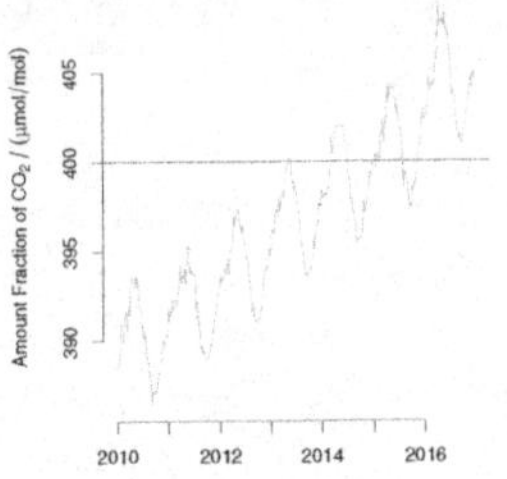

Weekly averages of determinations of the amount fraction of CO_2 in the atmosphere at the Mauna Loa Observatory (3397 m above sea level), in the Big Island of Hawaii, reported by NOAA's Global Monitoring Laboratory.

Since the 1950s, the Mauna Loa Observatory in the Big Island of Hawaii has made measurements of different gases in the atmosphere, most notably of carbon dioxide. The graph alongside, known as the "Keeling Curve" in honor of the late Charles David Keeling (1928–2005), shows the relentless rise of CO_2 levels whose pattern extends all the way back to the onset of the observations. The weekly averages generally increase approximately linearly throughout 2010–2016, while exhibiting a marked seasonal oscillation.

CO_2 levels decrease throughout the summer when plants take it during photosynthesis, and increase starting in the fall when decomposing plant matter releases it back into the atmosphere.

An autoregressive, integrated moving-average (ARIMA) model [Box et al., 2008] with linear drift, and with one auto-regressive term and one moving average term, together with a first order auto-regression for the first differences of the seasonal component with frequency of 52 weeks per year, provides an accurate fit to this time series. The model was chosen based on the Bayesian Information Criterion (BIC, Page 100), and was fitted to the series of 365 weekly averages using R function `Arima` defined in package `forecast`. [50] [51]

[50] R.J. Hyndman and Y. Khandakar. Automatic time series forecasting: the forecast package for R. *Journal of Statistical Software*, 27:1–22, July 2008

[51] R. J. Hyndman and G. Athanasopoulos. *Forecasting: Principles and Practice*. OTexts, Melbourne, Australia, second edition, 2018. ISBN 978-0-9875071-1-2. URL `http://OTexts.com/fpp2/`

The corresponding residuals, however, are not consistent with the assumption that the innovations in the ARIMA are Gaussian. For this reason, when we employ the parametric bootstrap below, we resample from these residuals directly, instead of from a Gaussian distribution with the same standard deviation that these residuals have.

Owing to the uncertainty surrounding the weekly averages, the fact that they exceeded the threshold of

400 µmol/mol for the first time during the week of May 9th, 2013, does not imply that the true amount fraction did, too. The parametric statistical bootstrap [Efron and Tibshirani, 1993] can be employed to estimate and characterize the uncertainty of the epoch at which the true amount fraction will have exceeded that threshold.

Comparisons between values of the amount fraction measured for flask samples in the laboratory, and values measured *in situ* at the Mauna Loa Observatory, reveal that the latter are slightly biased downward, being too small by about 0.15 µmol/mol on average.[52]

[52] K. W. Thoning, P. P. Tans, and W. D. Komhyr. Atmospheric carbon dioxide at Mauna Loa Observatory: 2. Analysis of the NOAA GMCC data, 1974–1985. *Journal of Geophysical Research: Atmospheres*, 94(D6):8549–8565, 1989. doi:10.1029/JD094iD06p08549

These data are made available at `gml.noaa.gov/ccgg/about/co2_measurements.html` by Pieter Tans and Kirk Thoning (NOAA Global Monitoring Laboratory, Boulder, Colorado).

The standard deviation of the differences between the daily observations and the corresponding weekly averages, is around 0.4 µmol/mol. Even though weekly averages likely will be less variable than daily observations (by how much depends on the autocorrelation of the time series of daily observations), in the following R code we adopt this value as standard uncertainty for the measurement error, exclusive of the bias.

First, we read the series of measurements directly from the corresponding NOAA repository, select the segment we wish to model, 2010–2016, and fit the ARIMA model to them.

```
require(forecast); require(lubridate)

URL = paste0("https://gml.noaa.gov/webdata/ccgg/trends/",
             "co2/co2_weekly_mlo.txt")
co2 = read.table(url(URL), header=FALSE)
# select data from 2009-2017
co2 = co2[(co2[,1]>2009) & (co2[,1]<2017),c(1,2,3,5)]
names(co2) = c("year", "month", "day", "x")

co2$date = make_date(year=co2$year, month=co2$month, day=co2$day)
co2$week = week(co2$date)
co2.ts = ts(co2$x, start=c(2010, 1), frequency=52)

co2.arima = Arima(co2.ts, order=c(1,0,1), seasonal=c(1,1,0),
                  include.drift=TRUE)
```

Second, we employ a Monte Carlo procedure to account for the bias as a fixed effect, take into account the measurement error aforementioned (which we model as being Gaussian), and we also resample the residuals from the original seasonal ARIMA model, which we use as additional perturbations for the original fitted values.

```
r = residuals(co2.arima)
n = length(r)
K = 1000
cross = rep(NA, K)
for (k in 1:K)
{
  y = fitted(co2.arima) + sample(r, size=n, replace=TRUE) +
      rnorm(n, mean=+0.15, sd=0.4)
  y.arima = try(Arima(y, order=c(1,0,1), seasonal=c(1,1,0),
                include.drift=TRUE))
  if (class(y.arima) == "try-error") { next }
  else { yHAT = fitted(y.arima)
         cross[k] = min((1:n)[yHAT > 400]) }
}
```

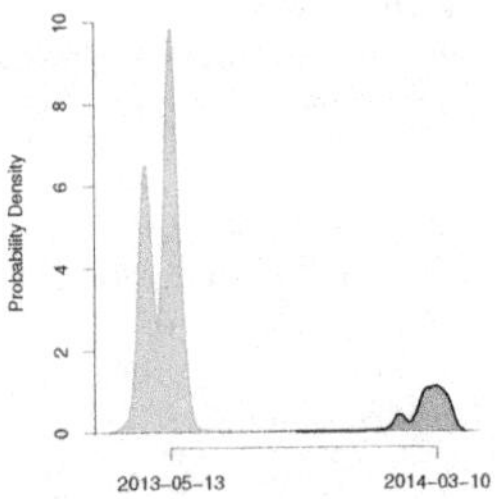

Week of first exceedance of the 400 µmol/mol threshold at the Mauna Loa Observatory, taking measurement uncertainty into account.

The first excursion of the true value of the amount fraction of atmospheric CO_2 above the 400 µmol/mol threshold may have occurred in May 2013 with 86 % probability, or much later, in March 2014, with 14 % probability. (These probabilities are the areas of the regions shaded light or dark gray under the probability density (Page 159) estimate depicted alongside.)

Comparing Two Measurement Methods

Laboratory practice often involves comparing a new or less-established method with an established standard method. The mass concentration of fat in human milk may be determined based on the measurement of glycerol released by enzymatic hydrolysis of triglycerides [Lucas et al., 1987], or by the Gerber method [Badertscher et al., 2007], which measures the fat directly with a butyrometer, after separating the fat from the proteins.

The following 45 pairs of values of the mass concentration of fat in human milk (expressed in cg/mL) were determined based on enzymatic hydrolysis of triglycerides (Trig), and by the Gerber method (G), as reported by Bland and Altman [1999, Table 3].

γ_{Trig}	γ_{G}	γ_{Trig}	γ_{G}	γ_{Trig}	γ_{G}	γ_{Trig}	γ_{G}
0.96	0.85	1.93	1.88	2.67	2.70	4.20	4.27
1.16	1.00	1.99	2.00	2.61	2.70	4.05	4.30
0.97	1.00	2.01	2.05	3.01	3.00	4.30	4.35
1.01	1.00	2.28	2.17	2.93	3.02	4.74	4.75
1.25	1.20	2.15	2.20	3.18	3.03	4.71	4.79
1.22	1.20	2.29	2.28	3.18	3.11	4.71	4.80
1.46	1.38	2.45	2.43	3.19	3.15	4.74	4.80
1.66	1.65	2.40	2.55	3.12	3.15	5.23	5.42
1.75	1.68	2.79	2.60	3.33	3.40	6.21	6.20
1.72	1.70	2.77	2.65	3.51	3.42		
1.67	1.70	2.64	2.67	3.66	3.62		
1.67	1.70	2.73	2.70	3.95	3.95		

The correlation coefficient for these two sets of measured values is quite high, 0.998, but it is a misleading indication of agreement between two measurement methods because a perfect correlation only indicates that the value measured by one method is a linear function of the value measured by the other, not that the corresponding measured values are in close agreement.

[53] B. Carstensen. *Comparing Clinical Measurement Methods*. John Wiley & Sons, Chichester, UK, 2010

The results of the paired t-test indicate that the mean difference does not differ significantly from zero.[53] However, this, too, falls short of establishing equivalence (or, interchangeability) between the two measurement methods. If the paired samples are of small size, then there is a fair chance that a statistical test will fail to detect a difference that is important in practice. And if they are of a large size, then a statistical test very likely will deem significant a difference that is irrelevant in practice.

Bland and Altman [1986] is the most often cited article in the *Lancet*, which reveals the exceptional interest that measurement issues enjoy in medicine. In 2014, *Nature* recognized this article as the 29th most-cited research of all time, over all fields.

For these reasons, Bland and Altman [1986] suggest that the question of agreement between methods be answered using suitable graphical methods.

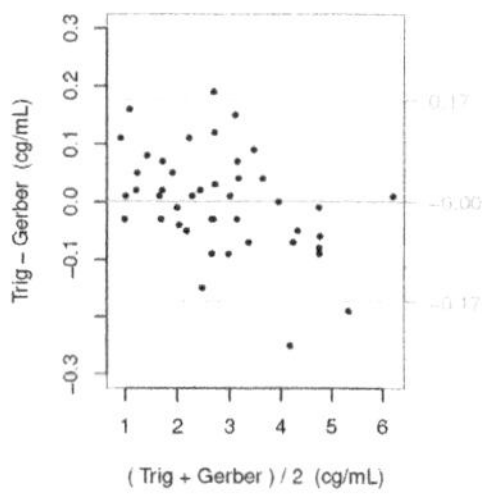

Bland-Altman plot, with the *limits of agreement* at ±0.17 cg/mL.

The Bland-Altman plot shows how the difference between the paired measured values varies with their averages [Altman and Bland, 1983; Bland and Altman, 1986]. Except for the inclusion of *limits of agreement* (the average of the differences between paired measured values plus or minus twice the standard deviation of the same differences), the Bland-Altman plot is similar to Tukey's mean-difference plot.[54]

[54] J. Chambers, W. Cleveland, B. Kleiner, and P. Tukey. *Graphical Methods for Data Analysis*. Wadsworth, Belmont, CA, 1983

In this case, the difference between the methods tends to be positive for small values of the measurand, and negative for large values. This feature can be illustrated using a variant of the Bland-Altman plot that recognizes such trend. Function `BA.plot` from R package `MethComp` [Carstensen et al., 2020] was used to draw the Bland-Altman plots.

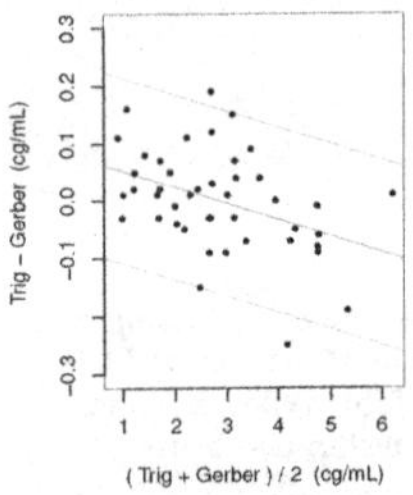

Bland-Altman plot recognizing that the differences between paired measured values depend on the averages of the same values.

Two methods are commonly employed to obtain the linear equation that "converts" a value produced by the Gerber method into the value that Trig would be expected to produce: the so-called *Deming regression* [Deming, 1943] and *Passing-Bablok regression* [Passing and Bablok, 1983].

These two regression lines can be computed using R functions defined in package `MethComp` as follows:

```
require(MethComp)
Deming(x=Gerber, y=Trig, vr = 1, boot=TRUE)
PBreg(x=Gerber, y=Trig)
```

Deming regression fits a straight line to points of a scatterplot when both coordinates are measured with error (ordinary linear regression assumes that only the response variable is measured with error). It assumes that both variables are affected by measurement uncertainty and the ratio of these variances, λ, is assumed to be known (in this case, $\lambda = 1$ is assumed). Thus, we have the following statistical model:

$$\gamma_{\mathrm{G},i} \sim \mathrm{GAU}(\xi_i,\ \sigma^2),$$
$$\gamma_{\mathrm{Trig},i} \sim \mathrm{GAU}(a + b\xi_i,\ \lambda\sigma^2).$$

The likelihood function for this problem is therefore a function of a, b, $\{\xi_i\}$, and σ^2 and the maximum likelihood estimates of these parameters can be obtained analytically.

Passing-Bablok regression estimates the coefficients a and b in

$$\gamma_{\text{Trig}} = a + b\gamma_{\text{G}}$$

as follows: the slope b is the median of the slopes of the straight lines between every pair of points (excluding any resulting slopes that are either 0 or infinity), and the intercept a is the median of the intercepts $\{y_i - bx_i\}$ determined by each of the points.

In this case, these methods yield the following lines:

The slope is consistent with the fact that only about 98 % of the fat in human milk is present as triglycerides [Lucas et al., 1987], which are the target of Trig.

$$\begin{aligned} \text{Deming:}\quad \gamma_{\text{Trig}} &= 0.078 + 0.972 \times \gamma_{\text{Gerber}}, \\ \text{Passing-Bablok:}\quad \gamma_{\text{Trig}} &= 0.055 + 0.976 \times \gamma_{\text{Gerber}}. \end{aligned}$$

With 95 % confidence, the true slopes are believed to lie in these intervals:

$$\begin{aligned} \text{Deming Slope: } &[0.953, 0.988], \\ \text{Passing-Bablok Slope: } &[0.956, 0.995]. \end{aligned}$$

Since these intervals exclude the equivalence value of 1.000, we can conclude that the two methods do not provide equivalent results.

To declare that two measurement methods are equivalent, not only should they produce results that are in agreement with due allowance for their respective uncertainties, over the relevant range of concentrations, but the measurement uncertainties that they typically achieve also should be in fair agreement.

Interpolating

Interpolation is a method of estimating values that might have been observed "in-between" values that actually were observed, where the "in-between" typically refers either to epochs in time or to locations in space. This is usually accomplished by determining a trend across the observations.

First, we consider linear interpolation of the pH of beer between values of pH for two standards, based on voltages generated in a pH meter observed for the standards and for the beer sample of interest.

Next we turn to the interpolation of temperature between fixed points of the International Temperature Scale of 1990 (ITS-90), to determine the temperature that corresponds to a reading produced by a platinum resistance thermometer at the freezing point of cadmium.

Finally, we employ Gaussian process regression to characterize the relationship between the chirping frequency of crickets and ambient temperature, which involves interpolation to estimate chirping frequency at temperatures that were not observed during the study.

pH of Beer

Determining the acidity of beer is important to control the brewing process and ensure consistent results. It was for this reason that Danish chemist Søren Sørensen, the Director of the Carlsberg Research Laboratory, introduced the concept of pH in 1909 and established the use of pH standards (buffers) in biochemistry.

To determine the pH of a beer sample B, readings of instrumental indications were obtained in quadruplicate, for the two standards (S_1 and S_2) and for the sample (B). The instrument used was a modern handheld pH meter.

Sample	pH	$u(\text{pH})$	E/mV	$u(E)$	ν
S_1	4.01	0.02	203.0	0.6	3
S_2	7.00	0.02	48.1	0.5	3
B			139.2	0.7	3

Each value of E in the foregoing table is an average of $m = 4$ such readings obtained under conditions of repeatability, $u(E)$ is the associated standard uncertainty evaluated by taking the standard deviation of the observations and diving them by $\sqrt{4}$, in accordance with the method described in the GUM 4.2.4. Therefore, each of these standard uncertainties is based on $\nu = 3$ degrees of freedom.

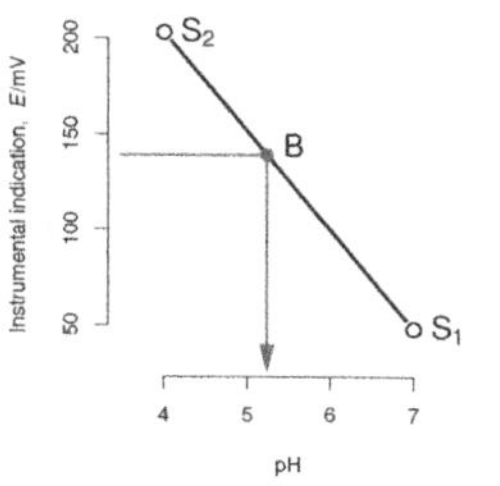

The pH of beer sample B that corresponds to the instrumental reading E_B is obtained by linear interpolation under the assumption that the instrumental indications vary linearly with the pH over the range of the two standards.

Finding the equation of a straight line that goes through points S_1 and S_2 amounts to solving these two equations for a and b:

$$E_1 = a + (b \times \text{pH}_1),$$
$$E_2 = a + (b \times \text{pH}_2).$$

Substituting the solutions into $E_B = a + (b \times \text{pH}_B)$ yields the following measurement model equation:

$$\text{pH}_B = \text{pH}_2 + \frac{\text{pH}_2 - \text{pH}_1}{E_2 - E_1}(E_B - E_2),$$

where pH_1 and pH_2 denote the pH of the two standards (S_1 and S_2), E_1 and E_2 are the corresponding average instrumental indications, and E_B is the average for the beer sample.

To evaluate $u(\text{pH}_B)$, one may use either Gauss's formula or a Monte Carlo method, similarly to how we evaluated the uncertainty associated with the volume of a storage tank. In both cases, the random variables used to model the reported uncertainties have standard deviations equal to the standard uncertainties.

When using Gauss's formula, the associated numbers of degrees of freedom are taken into account when assigning a Student's t distribution to the output quantity, pH_B,

	MODEL	MEAN	SD
pH_1	GAU	4.01	0.02
pH_2	GAU	7.00	0.02
E_1	t_3	203.0	0.6
E_2	t_3	48.1	0.5
E_B	t_3	139.2	0.7

as described in Annex G of the GUM [JCGM 100:2008]. For the Monte Carlo method, E_1, E_2, and E_B are modeled as Student's t_3 random variables, rescaled to have standard deviations $u(E_1)$, $u(E_2)$, and $u(E_B)$, respectively, and shifted to have means equal to the measured values of E_1, E_2, and E_B.

Since the relative uncertainties associated with the input quantities in the measurement model for pH_B all are quite small (none larger than 1 %), both approaches yield the same estimate $pH_B = 5.24$ and the associated uncertainty $u(pH_B) = 0.02$.

The R code below implements Gauss's method for this problem, which involves the computation of the first-order partial derivatives of pH_B with respect to each of the five input quantities.

```
## Measurement function
f = function(pH1, pH2, E1, E2, EB) {
              pH2 + (pH2 - pH1)*(EB - E2)/(E2 - E1) }

## Estimate of the measurand (GUM 4.1.4)
pHB = f(pH1=4.01, pH2=7.00, E1=203.0, E2=48.1, EB=139.2)

## Symbolic first-order partial derivatives of f
require(Deriv)
df = Deriv(f)

## Sensitivity Coefficients (GUM 5.1.3)
c = df(pH1=4.01, pH2=7.00, E1=203.0, E2=48.1, EB=139.2)

## Standard uncertainties associated with the input quantities
## and numbers of degrees of freedom they are based on
pH.u =  c(pH1=0.02, pH2=0.02, E1=0.6, E2=0.5, EB=0.7)
pH.nu = c(pH1=Inf,  pH2=Inf,  E1=3,   E2=3,   EB=3)

## Standard uncertainty associated with pHB (GUM 5.1.2)
pHB.u = sqrt(sum((c*pH.u)^2))

## Effective number of degrees of freedom (GUM G.4.1)
require(metRology)
pHB.nu = welch.satterthwaite(ui=pH.u, df=pH.nu, ci=c)

## Coverage interval with (GUM G.6.4)
pHB + qt(c(0.025, 0.975), df=pHB.nu) * pHB.u
```

One can also easily obtain the uncertainty budget that quantifies the influence that the uncertainties of the five input quantities have on $u(\mathrm{pH_B})$, similarly to Table H.1 in the GUM. This budget shows that the measurement of E_B makes the single largest contribution to $u^2(\mathrm{pH_B})$.

```
## Relative uncertainty contributions (%)
round(100*(c*pH.u)^2/sum((c*pH.u)^2), 1)
## pH1   pH2   E1    E2   EB
## 30.7  15.0  10.3  3.5  40.5
```

The R code below implements the Monte Carlo method for the same linear interpolation problem.

```
x    = c(pH1=4.01, pH2=7.00, E1=203.0, E2=48.1, EB=139.2)
x.u  = c(pH1=0.02, pH2=0.02, E1=0.6,   E2=0.5,  EB=0.7)
x.nu = c(pH1=Inf,  pH2=Inf,  E1=3,     E2=3,    EB=3)

K = 1e6
pH1 = rnorm(K, mean=x["pH1"], sd=x.u["pH1"])
pH2 = rnorm(K, mean=x["pH2"], sd=x.u["pH2"])
E1  = x["E1"] + x.u["E1"] * rt(K, df=x.nu["E1"]) /
                          sqrt(x.nu["E1"]/(x.nu["E1"]-2))
E2  = x["E2"] + x.u["E2"] * rt(K, df=x.nu["E2"]) /
                          sqrt(x.nu["E2"]/(x.nu["E2"]-2))
EB  = x["EB"] + x.u["EB"] * rt(K, df=x.nu["EB"]) /
                          sqrt(x.nu["EB"]/(x.nu["EB"]-2))
pHB = pH2 + (pH2 - pH1)*(EB - E2)/(E2 - E1)

plot(density(pHB))
c(pHB=mean(pHB), "u(pHB)"=sd(pHB),
  quantile(pHB, probs=c(0.025, 0.975)))
```

The probability distribution of $\mathrm{pH_B}$ has tails that are much heavier than Gaussian tails. According to the GUM, the distribution of $\mathrm{pH_B}$ should be approximately Student's t_{17}, rescaled to have standard deviation 0.02 and shifted to have mean 5.24, where the effective number of degrees of freedom, 17, was computed using the Welch-Satterthwaite formula (GUM Equation (G.2b)).

But the tails of $\mathrm{pH_B}$ are still much heavier than Student's t_{17}. Assuming that $\mathrm{pH_B}$ actually follows Student's t distribution as a working approximation, the maximum likelihood (Page 191) estimate derived from the Monte Carlo sample of $\mathrm{pH_B}$ suggests Student's t_5.

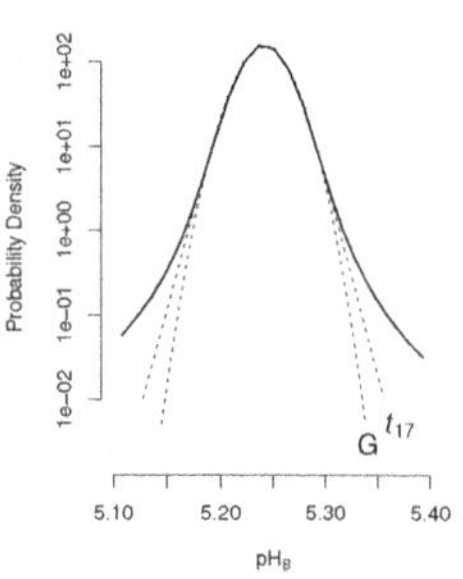

The probability distribution of $\mathrm{pH_B}$ (on a logarithmic scale) shows much heavier tails than Gaussian or Student's t_{17} distributions.

The measuring electrodes of the pH meters are sensitive to the pH but also to changes in temperature. Thus, when more precise pH measurements are needed, one has to take the temperature effect into account. This can be done by augmenting the equations that relate the instrumental readings and the pH:

$$E_1(T_1) = a + (b \times T_1 \times \mathrm{pH}_1(T_1)),$$
$$E_2(T_2) = a + (b \times T_2 \times \mathrm{pH}_2(T_2)).$$

As the above equations imply, in addition to the temperature dependence of E, the pH values of the standard solutions too depend on temperature and typically decrease by 0.01 – 0.02 pH units as the temperature of these standard solutions increases by 5 °C.

Freezing Point of Cadmium

The freezing point of a metal is the temperature at which the liquid metal becomes solid. Because the freezing points of many metals can be determined reliably, they are used as natural reference points for temperature. Indeed, the International Temperature Scale of 1990 assigns fixed temperature values to the freezing points of seven metals. When fixed points are realized, the temperature sensor should be close to, and surrounded by the interface between the liquid and the solid metal.[55]

[55] B. W. Mangum and G. T. Furukawa. *Guidelines for Realizing the International Temperature Scale of 1990 (ITS-90)*. National Institute of Standards and Technology, Gaithersburg, MD, 1990. NIST Technical Note 1265

The table below lists measurement results for averages of four replicate ratios of resistance values from Mangum et al. [2002, Table 1]. These are ratios between values of resistance measured using a Hart Model 5681 standard platinum resistance thermometer (SPRT), for three fixed points of the ITS-90 and for the freezing point of cadmium, and the resistance when the same SPRT is immersed in the triple point of water (TPW) cell:

In its most basic form, a platinum resistance thermometer consists of a long, thin platinum wire wrapped around a ceramic or glass core. The resistance of the wire, which changes *nearly* linearly with temperature, is used to provide an indication of temperature. Resistance is measured using a Wheatstone bridge circuit.

$$W(T_{\mathrm{FP}}) = \frac{R(T_{\mathrm{FP}})}{R(T_{\mathrm{TPW}})}.$$

The temperatures listed in the following table for the freezing points of Al, Zn, and Sn are assumed known with full certainty under the convention of ITS-90.

FP	T/K	$W(T_{\mathrm{FP}})/(\Omega/\Omega)$	$u(W(T_{\mathrm{FP}}))$
Al	933.473	3.375 732 56	0.000 000 30
Zn	692.677	2.568 747 52	0.000 000 31
Sn	505.078	1.892 712 29	0.000 000 23
Cd		2.219 019 54	0.000 000 39

The *Guide to the Realization of the* ITS-90,[56] on platinum resistance thermometry, specifies the steps that need to be taken to interpolate the temperature for the freezing point of cadmium.

[56] Consultative Committee for Thermometry. *Guide to the Realization of the ITS-90.* Bureau International des Poids et Mesures (BIPM), Sèvres, France, 2018. URL https://www.bipm.org/en/committees/cc/cct/guide-its90.html

First, we transform the measured resistance ratios of the three fixed points into their reference values using the ITS-90 reference function $W_{\mathrm{ref}} = f_{\mathrm{C}}(T)$. Then, we calculate the difference $\Delta W = W(T_{\mathrm{FP}}) - W_{\mathrm{ref}}(T_{\mathrm{FP}})$ and form a cubic interpolant:

$$\Delta W = a(W-1) + b(W-1)^2 + c(W-1)^3.$$

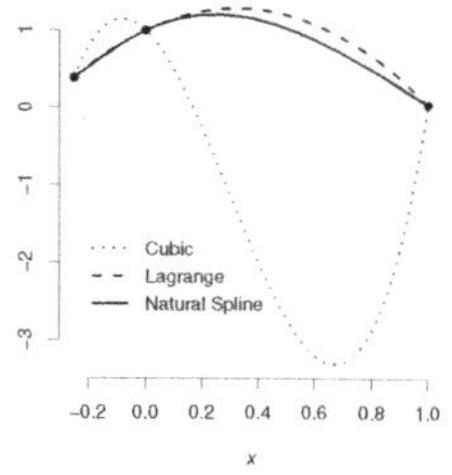

Both Lagrange polynomials and cubic natural splines are general purpose interpolants, not the particular cubic polynomial used for ΔW. In this illustration, the three points being interpolated have $x = -0.25, 0, 1$, and $y = 1/(1+25x^2)$.

Since there are three fixed points and the interpolant involves three coefficients (a, b, and c), the cubic fits them exactly.

It should be noted that this particular interpolant has been conceived for use in a very specific application, and is not a general purpose interpolant, as the figure alongside shows. In fact, Table 2 in the aforementioned guide specifies that this cubic interpolant should be used only for ratios of SPRT resistances within the range corresponding to the fixed points of Sn, Zn, and Al.

The interpolating cubic can now be used to calculate the deviation ΔW, hence W_{ref}, for any input value of W between the freezing points of Sn and Al. The temperature corresponding to the value of W_{ref} is obtained using the reference function $T = f_{\mathrm{D}}(W_{\mathrm{ref}})$, which is the inverse of the function f_{C} used earlier.

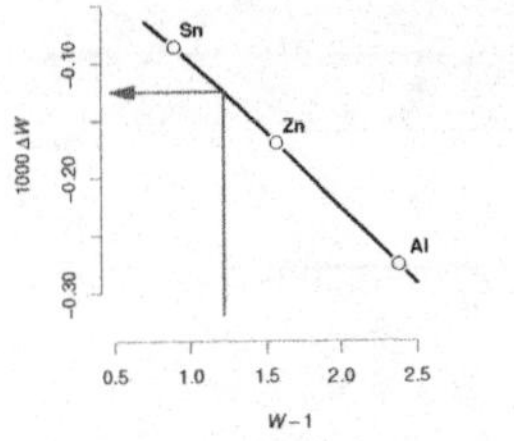

Interpolating the temperatures to determine the freezing point temperature of cadmium, amounts to following the arrow to obtain $W_{\text{ref}}(\text{Cd}) = W(\text{Cd}) - \Delta W_{\text{ref}}(\text{Cd})$ from $W(\text{Cd}) - 1$, which is then converted into temperature using the ITS-90 reference function, $f_D(W_{\text{ref}})$.

The following R code computes the interpolant based on the average resistance ratios W at the three ITS-90 fixed points, and then evaluates its mathematical inverse numerically at the average of the four replicates of W obtained at the freezing point of cadmium. We shall employ the parametric bootstrap to evaluate the uncertainty associated with the interpolated value of T_{Cd}.

```
Temp = c(Al=933.473, Zn=692.677, Sn=505.078)
W = c(Al=3.37573256, Zn=2.56874752, Sn=1.89271229)
uW = c(Al=0.00000030, Zn=0.00000031, Sn=0.00000023)
W.Cd = 2.21901954; uW.Cd = 0.00000039

# ITS-90 Reference functions
fC = function(Temp) {
  C = c(+1.64650916, -0.13714390, -0.00649767,
        -0.00234444, +0.00511868, +0.00187982,
        -0.00204472, -0.00046122, +0.00045724)
  2.78157254 + sum(C*((Temp - 754.15)/481)^(1:9)) }

fD = function(Wref) {
   D = c(+472.418020, +37.684494, +7.472018,
          +2.920828,  +0.005184, -0.963864,
          -0.188732,  +0.191203, +0.049025)
   273.15 + 439.932854 + sum(D*((Wref - 2.64)/1.64)^(1:9)) }
```

```
## Monte Carlo evaluation of Cd FP uncertainty
sBi = replicate(1e4, {
  WB = rnorm(3, mean=W, sd=uW * sqrt((4-1)/rchisq(3, 4-1)))
  ## Calculate (W - Wref) for Cd (y-axis)
  y = WB - c(fC(Temp[1]), fC(Temp[2]), fC(Temp[3]))
  ## Values for x-axis
  x = WB - 1
  ## Coefficients of the interpolating cubic spline
  sB = lm(y ~ 0 + x + I(x^2) + I(x^3))
  WCdB = rnorm(1, mean=W.Cd, sd=uW.Cd*sqrt((4-1)/rchisq(1,4-1)))
  ## Calculate delta W for Cd
  dWCd = predict(sB, newdata=data.frame(x=WCdB - 1))
  ## Use reference function to calculate T90 from Wref
  fD(yCd - dWCd)
  })
```

The International Temperature Scales ITS-1927 and ITS-1948 assigned a much lower value of 594.05 K to the freezing point of cadmium whereas Mangum et al. [2002] provide the modern estimate of it to be 594.22 K.

The resulting estimate of the freezing point temperature of cadmium is 594.2191 K, and a 95 % coverage interval for its true value ranges from 594.2187 K to 594.2195 K.

```
## Freezing point of Cd
c(mean(sBi), sd(sBi), quantile(sBi, probs=c(0.025, 0.975)))
```

Chirping Crickets

Gaussian process (GP) regression affords a more flexible way to model relationships between quantities than polynomial regression, which was used in the previous subsection to model the relationship between the resistance of a platinum wire and its temperature.

Neoconocephalus ensiger, a bush cricket. Photo by Marlo Perdicas (iNaturalist, Wikipedia).

Both GP regression and polynomial regression involve parameters whose values have to be estimated from the data. But while the values of the coefficients of a polynomial determine a particular polynomial, the values of the parameters of the GP determine the probability distribution of a random function.

A *random function* of temperature, say, is a collection of random variables indexed by values of temperature, X_{t_1}, ..., X_{t_m}, where the $\{t_i\}$ do not have to be equispaced. A collection of random variables indexed by a quantity (temperature in this case) for which some metric of "distance" is meaningful, is called a *stochastic process* [Hoel et al., 1972], or simply a *process*, for short.

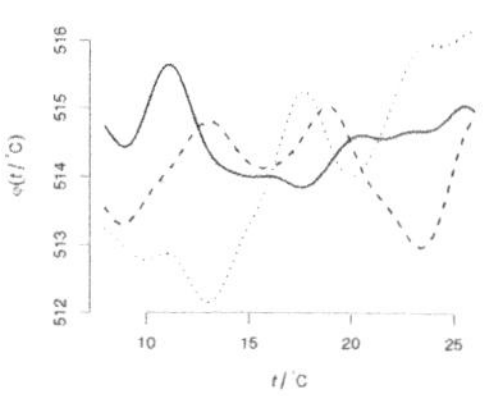

Three realizations of a Gaussian random function evaluated at 100 equispaced values of temperature. In each realization, the 100 values of φ are joined by straight line segments, creating the appearance of the graph of a continuous function.

As the figure alongside shows, random functions need not be wiggly and jagged, but can be very smooth. The smoothness is achieved by introducing correlations between these random variables. If $|t_i - t_j|$ is small, then X_{t_i} and X_{t_j} will be strongly correlated, but as $|t_i - t_j|$ increases, the correlation between them approaches zero.

The strong, short-range correlations induce nearby values to be similar, hence induce smoothness, while the long-distance correlations being close to zero give the function the freedom to oscillate considerably throughout the interval of values of temperature where it is defined. Gramacy [2020, Chapter 5] provides a very clear, accessible overview of GP regression.

In 1897, American physicist Amos Dolbear, better known for his inventions of telegraphs and telephones, wrote to *The American Naturalist* on the regularity of cricket chirps:[57]

[57] A. E. Dolbear. The cricket as a thermometer. *The American Naturalist*, 31(371):970–971, 1897. doi:10.2307/2453256

At night when great numbers are chirping the regularity is astonishing, for one may hear all the crickets in a field chirping synchronously, keeping time as if led by the wand of a conductor.

$t/°C$	ν/min^{-1}
8	264
9	285
14	346
17	417
18	438
19	495
20.5	524
21.5	540
23	643
24	693
25	744
26	780

Average number of chirps per minute from eleven male sword-bearing bush crickets for a series of temperatures.

The chirping rate seems to be determined entirely by ambient temperature but the relationship it not quite as simple as Dolbear put it. We shall use the results of measurements that Frings and Frings [1957] made in a laboratory experiment involving eleven male sword-bearing bush crickets, *Neoconocephalus ensiger*, whose chirps they recorded.

The GP regression model expresses the frequency of chirps as $\nu_t = x_t + \varepsilon_t$ where $\{x_t\}$ is a GP and $\{\varepsilon_t\}$ are independent, identically distributed Gaussian random variables with mean 0 and standard deviation τ. The GP tracks the smooth trend (thick black curve in the figure alongside) of chirping frequency as a function of temperature, while the $\{\varepsilon_t\}$ "explain" deviations from such trend, for example at 19 °C and at 21.5 °C.

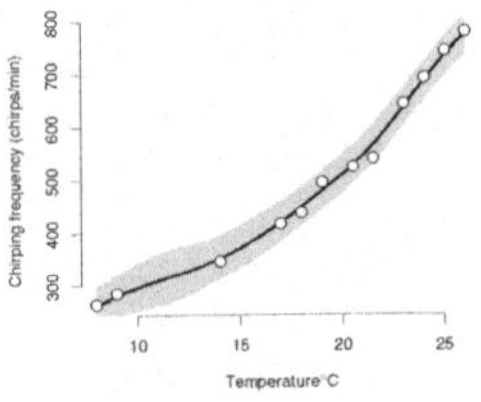

Gaussian Process regression fit to bush cricket chirping frequency data at various temperatures. The thick black line represents the mean of the GP, and the shaded band represents the associated uncertainty in the form of a 95 % coverage interval.

The model was fitted to these data using a Bayesian procedure implemented in R function bgp from package tgp [Gramacy, 2007], yielding the curve and surrounding uncertainty depicted alongside:

```
temp = c(8, 9, 14, 17, 18, 19, 20.5, 21.5, 23, 24, 25, 26)
freq = c(264, 285, 346, 417, 438, 495, 524, 540, 643,693,744,780)
require(tgp)
gp = bgp(X=temp, Z=freq, bprior="b0", corr="exp",
        BTE=c(10000, 100000, 25),
        XX=seq(from=min(temp), to=max(temp), length=500))
```

Function bgp carries out these four tasks:

(1) Estimates the parameters of the Gaussian Process regression model based on the twelve pairs of observations $\{(t_1, \nu_1), \ldots, (t_{12}, \nu_{12})\}$;

(2) Evaluates the uncertainty associated with these estimates;

(3) Uses the fitted model to predict values of the frequency ν at all the other values of temperature where the chirping frequency was not observed (these predictions are needed to be able to draw points along the trend sufficiently close to one another to create the semblance of a continuous curve); and

(4) Evaluates the uncertainty that surrounds the resulting Gaussian Process regression curve.

The centerpiece of the model is the correlation function, whose default option in bgp is the powered exponential such that $\rho(s) = \exp\{-(s/\phi)^{\alpha}\}$ is the correlation between X_t and X_{s+t} for $s > 0$, whose parameters are the scale $\phi > 0$, and shape $0 < \alpha \leqslant 2$.

This means that the correlation between chirping frequencies at different temperatures depends only on the difference between the temperatures and decays exponentially fast to zero as the difference between values of temperature increases.

The model includes also a scale parameter σ that, upon multiplication by $\rho(s)$, yields all the elements of the 12×12 covariance matrix for the observations of chirping frequency, and an overall mean μ that sets the typical level of this frequency.

The Bayesian approach is particularly helpful in this context because it recognizes and propagates the uncertainties associated with the estimates of all these parameters (Task (2) above) as part and parcel of the model fitting procedure, not as an add-on or afterthought.

Calibrating

When a truck stops at a highway scale to be weighed, it applies a force to one or more load cells (Page 103) under the scale, which generates a potential difference between the electrical terminals that the load cells are connected to. Calibration is the procedure that establishes a relation between values of the force applied to a load cell and corresponding values of potential difference, thereby making possible to "translate" indications of electric potential (voltage) into values of force. These values of force, in turn, are translated into values of mass using the local value of Earth's gravitational acceleration and Newton's second law of motion.

Calibrating a measuring instrument consists of determining a relationship between values of the measurand, and the typical, corresponding instrumental responses (or, *indications*), and characterizing the uncertainty surrounding such relationship. This is usually done by exposing the instrument to several different, known (up to measurement uncertainty) values of the measurand in measurement standards, making suitably replicated observations of the instrumental responses that these exposures generate, and finally deriving the typical responses from these observations.

The aforementioned relationship is often described by means of a *calibration function* that maps values of the measurand to typical (or, expected) values of the indications produced by the instrument being calibrated. For example, the result of calibrating a thermocouple for use as a thermometer is either a mathematical function that maps values of temperature into values of voltage, or a table that lists the values of voltage that correspond to specified values of temperature.

To be able to use the instrument to make measurements, the inverse relationship is needed, which produces an estimate of the value of the measurand given an observed instrumental response. This is variously called the *analysis function*, *measurement function*, or the *evaluation function*, depending on the field of application.

We begin by illustrating the development of calibration and analysis functions for the measurement of the mass concentration of chloromethane (Page 98) using gas chromatography and mass spectrometry, and in the process introduce criteria for model selection, and demonstrate Monte Carlo methods for uncertainty evaluation.

In this case, a very simple function, a cubic polynomial without the quadratic term, strikes just the right balance between goodness-of-fit to the calibration data and model simplicity. Many measurement systems, however, require calibration functions of much greater complexity.

For example, the calibration of capsule-type standard platinum resistance thermometers over the range 13.80 K (triple point of hydrogen) to 273.16 K (triple point of water) in NIST SRM 1750 involved determining a polynomial of the 7th degree to describe the deviations between the ITS-90 reference curve for this range, and the actual values of resistance for these resistance thermometers [58]. An even more complex model is often used to characterize the dose-response of many bioassays, involving a five-parameter logistic function.[59]

One of the most complex calibration models used currently in science involves a Bayesian spline model with consideration of errors-in-variables that serves to convert measurements of carbon-14 concentration into measurements of the age of a biological material, in a technique known as *radiocarbon dating* (Page 118).

[58] W. L. Tew and G. F. Strouse. *Standard Reference Material 1750: Standard Platinum Resistance Thermometers, 13.8033 K to 429.7485 K.* NIST Special Publication 260-139. National Institute of Standards and Technology, Gaithersburg, MD, November 2001. doi:10.6028/NIST.SP.260-139

[59] P. G. Gottschalk and J. R. Dunn. The five-parameter logistic: A characterization and comparison with the four-parameter logistic. *Analytical Biochemistry*, pages 54–65, 2005. doi:10.1016/j.ab.2005.04.035

Chloromethane

Chloromethane is a volatile organic compound with boiling point −24 °C at normal atmospheric pressure, and chemical formula CH_3Cl. It is currently used industrially as a reagent and solvent, and in the past was widely used as a refrigerant. Chloromethane is water-soluble and its concentration in water is usually measured using gas chromatography and mass spectrometry (GC-MS).[60]

[60] J. W. Munch. *Method 524.2. Measurement of Purgeable Organic Compounds in Water by Capillary Column Gas Chromatography/Mass Spectrometry*. National Exposure Research Laboratory, Office of Research and Development, U.S. Environmental Protection Agency, Cincinnati, OH, 1995. Revision 4.1

The table below lists replicated instrumental indications obtained with a GC-MS system to measure mass concentration of chloromethane, using fluorobenzene as internal standard [Lavagnini and Magno, 2007]: the indications are ratios between areas of peaks in the traces produced by the measuring system, one corresponding to chloromethane, the other corresponding to a known amount of the internal standard that is injected into the system simultaneously with each sample of each chloromethane standard, thereby correcting for losses of the measurand (or, analyte) in the standard as it travels through the GC column.

Concentration of chloromethane, c (μg/L)								
0.00	0.03	0.10	0.20	0.40	0.80	1.60	3.20	4.00
9219	12 867	24 122	36 817	51 036	111 975	174 220	344 967	355 100
9101	12 675	20 211	38 457	53 503	84 405	172 282	297 678	341 706
6914	14 311	20 900	31 085	64 271	95 427	168 291	308 669	365 223
8310	12 292	20 327	36 355	55 831	118 919	152 625	277 519	363 193
7603	9007	23 622	44 505	71 737	125 506	229 081	351 525	417 577
9011	11 415	19 576	37 588	57 600	89 315	216 992	302 684	389 765
6061	14 701	26 155	30 706	75 693	116 848	186 974	389 644	411 681
8032	13 757	18 471	34 256	66 599	138 121	176 933	323 136	390 485
5932	12 900	30 002	37 076	59 649	126 417	242 466	358 242	465 813
6034	12 800	29 385	42 269	64 498	105 840	239 470	366 867	444 202

The entries in the body of the table are values of $r \times 10^6$

For each of nine $CHCl_3$ calibration standards, ten replicate measurements of the ratio r of areas of peaks produced by the GC-MS measuring system, that correspond to $CHCl_3$ and to the internal standard [Lavagnini and Magno, 2007, Table 2].

A plot of the values of r against corresponding values of c shows that the dispersion of the replicated values of r increases substantially with increasing values of c. This undesirable feature is much reduced once the data are re-expressed [Mosteller and Tukey, 1977, Chapters 4-6] using logarithmic scales, which also implies that the focus is on the relative uncertainties.

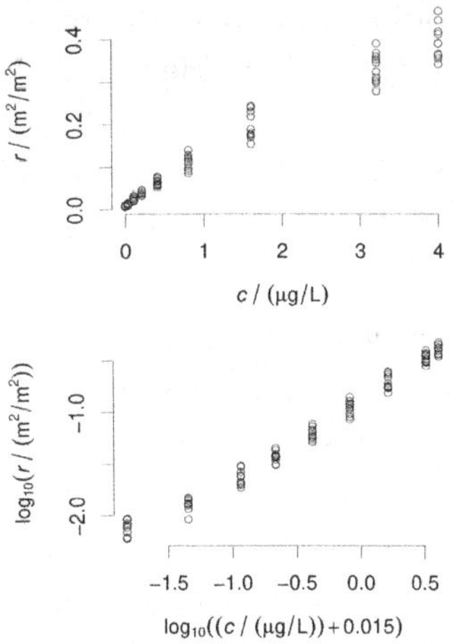

Calibration data before and after re-expression using logarithms. The addition of 0.015 serves to avoid taking logarithms of zero, but otherwise it is inconsequential.

We will neglect the uncertainty surrounding the values of c because, in this particular case, in fact it is negligible by comparison with the dispersion of the replicated values of r. (Possolo [2015, E17] describes an instance of calibration where uncertainties surrounding the values of the measurand in the calibration standards, and the instrumental indications, both have to be recognized.)

Model selection is the task of choosing a model to represent how $R = \log_{10}(r)$ varies as a function of

$$C = \log_{10}(c/(\mu g/L) + 0.015).$$

Several polynomial models may be used to summarize the relationship between them. For example,

$$R = \alpha + \beta_1 C,$$
$$R = \alpha + \beta_1 C + \beta_2 C^2 + \beta_3 C^3, \text{or}$$
$$R = \alpha + \beta_1 C + \beta_3 C^3,$$

because one may either add or remove terms while searching for the best model. As more and more terms involving different powers of C are added to the model, the polynomial fits the data ever more closely. When to stop, and which model to choose?

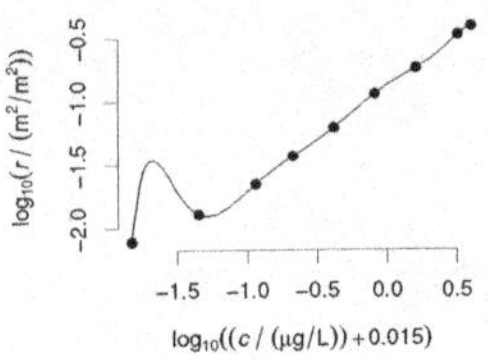

A polynomial may fit the data exactly and still be an awful calibration function.

Suppose we would summarize the replicated values of r that correspond to each value of c with their median, and fitted a polynomial of the 8th degree to these nine points. This polynomial fits the summary data exactly, but look how it behaves around the two leftmost points!

While inappropriate here, polynomials of high degree are used occasionally as models. The International Temperature Scale ITS-90, for example, uses polynomials of the 9th and 15th order as reference functions.

The goal of *model building* is to achieve a representation of the data that is accurate enough for the purpose the model is intended to serve, while keeping the model as parsimonious as possible. Parsimony, in this case, means small number of adjustable parameters, or low degree of the polynomial. The reason why parsimony matters is that simple models generally have better real-world performance than extravagant models.

For a polynomial model, fitting the model to the data amounts to finding values of the coefficients that make the graph of the polynomial pass as closely as possible to the data points. Several aspects of this issue are discussed under *Least Squares* (Page 196).

MODEL, φ	BIC(φ)
$\alpha + \beta_1 C$	−190
$\alpha + \beta_1 C + \beta_2 C^2$	−226
$\alpha + \beta_1 C + \beta_2 C^2 + \beta_3 C^3$	−231
$\boldsymbol{\alpha + \beta_1 C + \beta_3 C^3}$	**−235**
$\alpha + \beta_1 C + \beta_2 C^2 + \beta_3 C^3 + \beta_4 C^4$	−227
$\alpha + \beta_1 C + \beta_2 C^2 + \beta_3 C^3 + \beta_4 C^4 + \beta_5 C^5$	−222

The smaller the value of the Bayesian Information Criterion, BIC, the more adequate is the model for the data. In general, a difference in BIC values greater than 10 is strong evidence against the model with the higher BIC value, whereas a difference of less than 2 is considered insignificant. Thus, and in this case, the models in the third and fourth rows of this table are comparably adequate for the data.

A reliable guide for model building aims to strike a compromise between goodness of fit and simplicity. One such guide is the *Bayesian Information Criterion* (BIC) [Burnham and Anderson, 2004], which we explain in *Model Selection* (Page 201). For now, it suffices to say that the smaller the BIC, the more adequate is the model.

For the GC-MS calibration data listed above, the best model happens to be a polynomial of the third degree without the quadratic term,

$$\varphi(C) = \alpha + \beta_1 C + \beta_3 C^3,$$

with coefficients $\widehat{\alpha} = -0.8931$, $\widehat{\beta}_1 = 0.8327$, $\widehat{\beta}_3 = -0.0473$. This defines the *calibration function*, which characterizes how the GC-MS instrument responds when exposed to standard solutions of chloromethane.

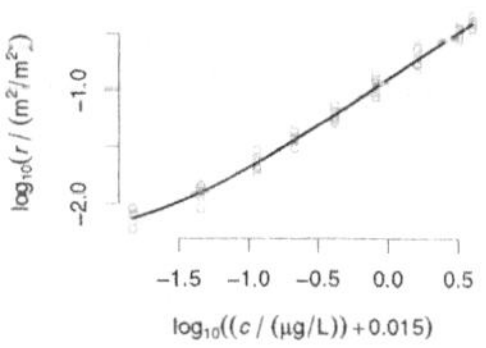

Calibration function, whose graph is the smooth curve, is a polynomial of the third degree without the quadratic term.

The analysis function is the mathematical inverse of the calibration function: ψ such that $\psi(\varphi(C)) = C$, for each value of C at which φ is defined. The analysis function is used to assign values of the measurand to samples whose mass concentration c of chloromethane is unknown, and which, upon injection into the GC-MS measuring instrument, produce a value of the ratio r.

Depending on the mathematical form of the calibration function φ, it may or may not be possible to derive an analytical expression (that is, a formula) for the analysis function ψ. However, it is always possible to determine it numerically given an observed value of R, by finding the values of C such that $\varphi(C) = R$. In case this equation is satisfied by more than one value of C, then some additional criterion needs to be employed to determine the appropriate solution: for example, the appropriate solution should lie between the minimum and maximum of the values of c in the standards used for calibration.

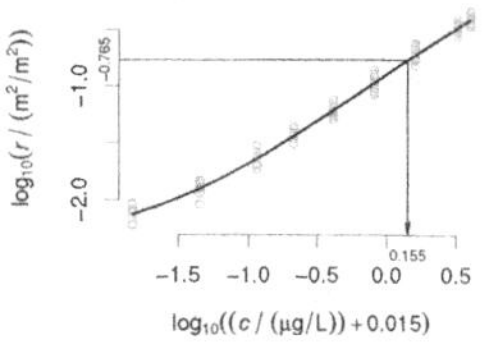

Determination of the value of c that corresponds to an instrumental indication $r = 0.1718\,\mathrm{m^2/m^2}$. Inversion of the calibration function produces $\log_{10}((c/\mu g/L) + 0.015) = 0.155$, hence $c = 1.41\,\mu g/L$.

The inversion that leads from φ to ψ can be performed without any mathematical derivations or numerical computation at all. First, draw the graph of the calibration function φ on a sheet of transparent acetate, with the horizontal axis indicating values of c increasing from left to right, and with the vertical axis indicating values of r increasing from bottom to top. Then, flip the sheet and look at it from the back side, and rotate it so that the vertical axis is now with values of c increasing from bottom to top, and horizontal axis with values of r increasing from left to right. The resulting graph depicts the analysis function ψ.

In the particular case under consideration, the calibration function is a polynomial of the third degree, and indeed it is possible to solve $\varphi(C) = R$ analytically for

C using a celebrated formula published in 1545 by Gerolamo Cardano, which implements the solution derived by Scipione del Ferro.

In practice, however, even in cases like this, solving the equation numerically may be the more expeditious route, allowing that most of the effort be dedicated to the selection of the most appropriate solution among the several that typically are available when the calibration function is a polynomial. This is how the graph of ψ was constructed that is displayed alongside: by solving $\varphi(C) = R$ for C for many equispaced values of R.

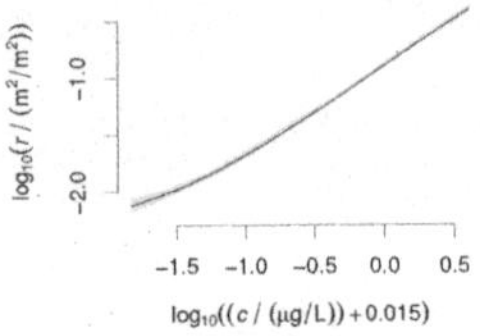

Calibration function φ and 95 % coverage band derived from 50 000 bootstrap replicates of this function.

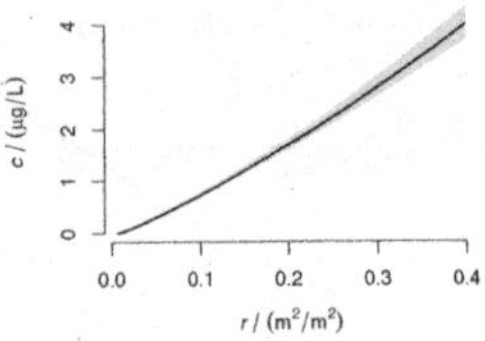

Analysis function ψ and 95 % coverage band corresponding to the calibration function above.

The evaluation of the uncertainty surrounding the calibration and analysis functions may be performed using a Monte Carlo method, which in this case will be the non-parametric statistical bootstrap invented by Bradley Efron and explained to perfection by Diaconis and Efron [1983]. The uncertainty evaluation is based on the results from many repetitions of these two steps:

(1) Draw a sample of size 90 from the set of 90 pairs $\{(c_{ij}, r_{ij})\}$ listed in the foregoing table, uniformly at random, with replacement: this means that all pairs have the same probability of being selected, and that each pair may be selected more than once;

(2) Use this sample as if it were the original data, and select and build a calibration function as described above — this is called a *bootstrap replicate* of the calibration function.

Each time these two steps are repeated yields a version of the calibration function φ. A band that contains 95 % of the graphs of the resulting versions of the calibration function, is a coverage band for the true calibration curve, as depicted alongside. The corresponding coverage band for the analysis function ψ is obtained by mathematical inversion of the upper and lower boundaries of the band that surrounds the calibration function.

Force Transducers

Honeywell model 3156 load cell to measure forces of up to 750 kN either under tension or compression.

A load cell is a force transducer – an electro-mechanical sensor that generates an electrical signal in response to the applied force. The electrical signal reflects a change in the resistance of a resistor strained by the applied force. This change in resistance is evaluated using a Wheatstone bridge (Page 24), and the electrical signal is expressed as a ratio between the load cell's output voltage and the voltage applied to it.

NIST uses deadweight machines to calibrate load cells. The following measurement results were obtained using the 4.45 MN machine whose stack of weights are depicted alongside. Figure 9 of Jabbour and Yaniv (2001) shows how the weight of one or more disks in the stack can be applied to the load cell, either in tension or in compression regimens.[61]

Stack of stainless steel weights in the 4.45 meganewton (approximately 454 000 kilogram-force) deadweight machine that the NIST uses to calibrate load cells.

The relative uncertainties associated with the responses are approximately twice as large as their counterparts for the forces because they reflect the appreciable sensitivity of this particular load cell to its placement in the machine that applies the forces. This uncertainty component has to be taken into account because NIST has no foreknowledge of how the owner of the load cell being calibrated will deploy it for use.

Characterizing a load cell means describing how it responds to applied forces, which is done by building a calibration function, φ, that maps values of the force, F, to values of the load cell's response, $R = \varphi(F)$. To use a calibrated load cell to measure force in practice one needs the mathematical inverse of the calibration function, $\psi = \varphi^{-1}$ such that $\psi(\varphi(F)) = F$. Bartel et al. [2016] call ψ the *measurement function* because it produces the value $F = \psi(R)$ of the force that corresponds to a response R produced by the load cell.

[61] Z. L. Jabbour and S. L. Yaniv. The kilogram and measurements of mass and force. *Journal of Research of the National Institute of Standards and Technology*, 106(1):25–46, January–February 2001

Differently from the calibration of the GC-MS instrument used to measure the concentration of chloromethane

(Page 98), here both quantities involved, the forces and the load cell's responses, are surrounded by uncertainty.

The calibration function, φ, relates the true values of the responses, $\{\rho_j\}$, to the true values of the applied forces, $\{\phi_j\}$, as $\rho_j = \varphi(\phi_j)$, for $j = 1, \dots,$ where, in this case, the number of calibration points is $n = 11$. Typically, φ is a low-degree polynomial, most commonly of the second or third degree, and never of degree greater than 5 for conformity with ASTM E74-13a, which is the documentary standard requested for most force transducers submitted to NIST for calibration.[62]

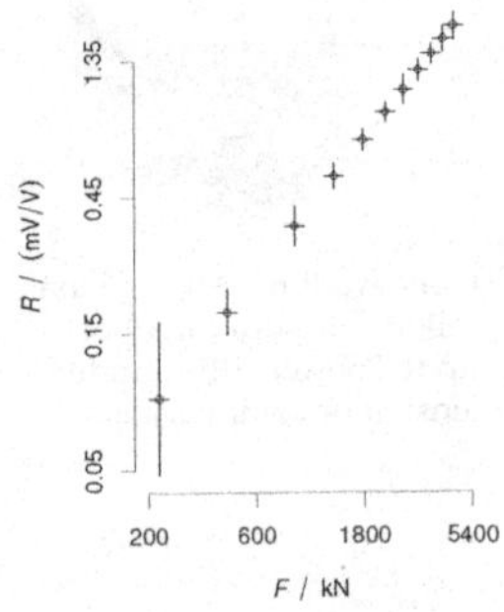

Measurement results, which are listed below in the block of R code used for model fitting, for the calibration of a load cell. The horizontal line segments represent $\{F_j \pm k_F u(F_j)\}$ and the vertical line segments represent $\{R_j \pm k_R u(R_j)\}$, where the magnification factors, $k_F = 20\,000$ and $k_R = 10\,000$, serve only to facilitate the visualization of the uncertainties, and are not used in the calculations that produce the calibration function. Notice that both axes have logarithmic scales.

Traditionally, the coefficients of the polynomial have been determined via ordinary least squares regression, which assumes that all substantively significant components of uncertainty are associated with the responses, $\{R_j\}$, and that the applied forces, $\{F_j\}$, are known with negligible uncertainty.

In fact, a conservative evaluation places the relative standard uncertainty in the measurement of the forces at 0.0005 %,[63] while the relative uncertainties in the measurement of the responses are much larger. However, recent advances in transducer technology challenge this assumption, so much so that the uncertainty surrounding the forces can contribute about 50 % to the uncertainty associated with the calibration function.

The uncertainty associated with the forces comprises uncertainty contributions from the determination of the mass of the deadweights, from the Earth's gravitational acceleration at the site where the machine is located, including its vertical gradient (which matters owing to the height of the stack of deadweights), and from the density of the air in the room where the machine is located (which impacts the buoyancy correction for the applied forces) [Bartel, 2005].

The uncertainty associated with the responses comprises uncertainty contributions from the lack of repeatability

[62] ASTM. *ASTM E74-13a, Practice of Calibration of Force-Measuring Instruments for Verifying the Force Indication of Testing Machines.* ASTM International, West Conshohocken, PA, 2013. doi:10.1520/E0074-13A

[63] T. Bartel. Uncertainty in NIST force measurements. *Journal of Research of the National Institute of Standards and Technology*, 110(6):589–603, 2005

of the transducer response indicating device, from the calibration of the instrumentation used to acquire the responses, and from the orientation of the transducer relative to the loading platens of the machine that applies the forces. For some load cells, including the one used for this illustration, the sensitivity to loading geometry can make the largest contribution to the $\{u(R_j)\}$ by far.[64]

To build the calibration function we will employ *errors-in-variables* (EIV) regression,[65] to fit the following model to the measurement results:

$$F_j = \phi_j + \varepsilon_j, \quad R_j = \rho_j + \delta_j, \quad \rho_j = \beta_1 + \beta_2\phi_j + \beta_3\phi_j^2,$$

for $j = 1, \ldots, n$, assuming that the measurement errors, $\{\varepsilon_j\}$, affecting the forces, are outcomes of independent Gaussian random variables all with the same mean 0 kN, and with standard deviations $\{u(F_j)\}$, and similarly for the measurement errors, $\{\delta_j\}$, affecting the responses.

A polynomial of the second degree was selected based on the Bayesian Information Criterion (BIC, Page 100), after comparing first, second, and third degree polynomials fitted by ordinary least squares.

The EIV model can be fitted to the calibration data either via maximum likelihood estimation (Page 193) or using a Bayesian procedure (Page 208). The two methods produce very similar estimates in this case, but the associated uncertainties are appreciably larger using maximum likelihood than using the Bayesian procedure defined using the Stan code listed below.[66]

	MLE		BAYES	
	ESTIMATE	STD. UNC.	ESTIMATE	STD. UNC.
β_1	−7.815 35	0.000 82	−7.815 36	0.000 39
β_2	0.997 233 7	0.000 220 5	0.997 234 0	0.000 106 7
β_3	0.000 194 97	0.000 014 83	0.000 194 95	0.000 007 20

The regression coefficients and associated uncertainties listed in the previous table are for the model fitted to

64 T. Bartel, S. Stoudt, and A. Possolo. Force calibration using errors-in-variables regression and Monte Carlo uncertainty evaluation. *Metrologia*, 53(3):965–980, 2016. doi:10.1088/0026-1394/53/3/965

65 R. J. Carroll, D. Ruppert, L. A. Stefanski, and C. M. Crainiceanu. *Measurement Error in Nonlinear Models — A Modern Perspective*. Chapman & Hall/CRC, Boca Raton, Florida, second edition, 2006

66 B. Carpenter, A. Gelman, M. Hoffman, D. Lee, B. Goodrich, M. Betancourt, M. Brubaker, J. Guo, P. Li, and A. Riddell. Stan: A probabilistic programming language. *Journal of Statistical Software*, 76(1):1–32, 2017. doi:10.18637/jss.v076.i01

the logarithms of the forces and of the responses, for the reasons that are explained below, which is of the form $\ln R_j = \beta_1 + \beta_2 \ln F_j + \beta_3 (\ln F_j)^2$ for $j = 1, \ldots, n$.

```
data {
  int n; // Number of observations
  vector<lower=0>[n] f;  // Force (N)
  vector<lower=0>[n] uf; // Std. unc. for force (N)
  vector[n] r;           // Response (mV/V)
  vector[n] ur;          // Std. unc. for response (mV/V)
  vector[3] betaMean;    // Prior mean for beta
  vector[3] betaSD;      // Prior SD for beta
  vector<lower=0>[n] phiMean; // Prior mean for phi (N)
  vector<lower=0>[n] phiSD;   // Prior SD for phi (N)
}
parameters {
  vector[3] beta;          // Regression coefficients
  vector<lower=0>[n] phi; // True values of force
}
transformed parameters {
  vector[n] rho; // True values of response
  for (j in 1:n) { rho[j] = beta[1] + beta[2]*phi[j]
                         + beta[3]*phi[j]^2;};
}
model {
  beta ~ normal(betaMean, betaSD); // Prior for beta
  phi ~ normal(phiMean, phiSD);    // Prior for phi
  f ~ normal(phi, uf);             // Likelihood for f
  r ~ normal(rho, ur);             // Likelihood for r
}
```

The model was fitted to the logarithms of the forces and of the responses, which is equivalent to focusing on the relative standard uncertainties instead of on the standard uncertainties. The resulting stabilization of the uncertainties (of the logarithms of the forces and responses) across the vast range of forces used during calibration facilitates MCMC sampling.

Note that Gauss's formula (Page 22) yields

$$u(\ln F_j) \approx u(F_j)/F_j, \text{ and}$$
$$u(\ln R_j) \approx u(R_j)/R_j, \text{ for } j = 1, \ldots, n.$$

Since the relative uncertainties for the responses are about twice as large as their counterparts for the forces, the weighted least squares estimates of the regression coefficients should provide a good approximation to their EIV counterparts, hence they are used to calibrate the prior distribution for these coefficients.

The following R code executes the previous Stan code after it will have been assigned to variable `eivModel`. R function `envelope` from the package `boot`[67] [68] was used to compute the 95 % coverage band for the true calibration function depicted in the adjoining figure.

[67] A. Canty and B. D. Ripley. *boot: Bootstrap R (S-Plus) Functions*, 2021. URL `cran.r-project.org/web/packages/boot/`. R package version 1.3-28

[68] A. C. Davison and D. Hinkley. *Bootstrap Methods and their Applications*. Cambridge University Press, Cambridge, UK, 1997. ISBN 0-521-57471-4. URL `statwww.epfl.ch/davison/BMA/`

```
z = read.table(header=TRUE, text="
          f      uf          r       ur
     ##        / kN              / (mV/V)
     222.4111 0.0011 0.0889000 5.50e-06
     444.8222 0.0022 0.1777733 3.30e-06
     889.6444 0.0044 0.3554480 5.80e-06
    1334.4666 0.0067 0.5331937 5.50e-06
    1779.2888 0.0089 0.7109277 6.40e-06
    2224.1110 0.0111 0.8887247 7.00e-06
    2668.9332 0.0133 1.0665180 1.28e-05
    3113.7554 0.0156 1.2443300 1.15e-05
    3558.5776 0.0178 1.4221563 1.21e-05
    4003.3998 0.0200 1.6000087 1.60e-05
    4448.2220 0.0222 1.7778847 2.00e-05")

## Model fitted to the logarithms of the forces and
## to the logarithms of the load cell responses
rl = log(z$r); url = z$ur/z$r
fl = log(z$f); ufl = z$uf/z$f
n = length(rl)

require(rstan)
options(mc.cores = parallel::detectCores())
rstan_options(auto_write = TRUE)

s = summary(lm(rl~poly(fl, degree=2, raw=TRUE), weights=1/url^2))

eivModel.data = list(n=n, f=fl, uf=ufl, r=rl, ur=url,
                     betaMean=s$coefficients[,"Estimate"],
                     betaSD=2*s$coefficients[,"Std. Error"],
                     phiMean=fl, phiSD=2*ufl)

eivModel.inits = function () {
    list(beta=rnorm(3, mean=s$coefficients[,"Estimate"],
                    sd=s$coefficients[,"Std. Error"]),
         phi=rnorm(n, mean=fl, sd=ufl)) }

eivModel.fit = stan(model_code = eivModel, data = eivModel.data,
                    control=list(adapt_delta=0.99,
                                 max_treedepth=20),
                    init=eivModel.inits,
                    warmup=100000, iter=250000,
                    chains=4, cores=4, thin=10)

print(eivModel.fit, digits=5)
eivModel.post = rstan::extract(eivModel.fit)
```

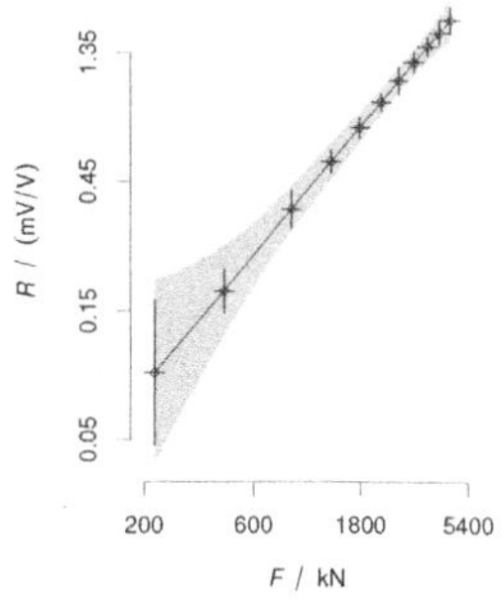

Calibration function (sloping line), and 95 % coverage band for its true value. The uncertainties for the forces are magnified 20 000 times, and the uncertainties for the responses are magnified 10 000 times. The scales of both axes are logarithmic.

Isochron for the Sea of Tranquility

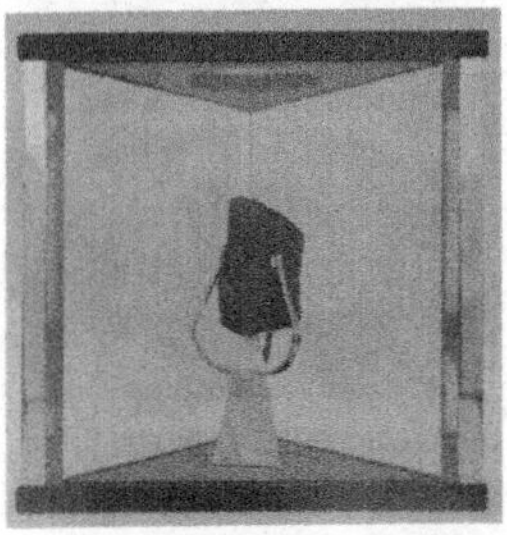

Largest fragment of moon rock 10057 collected during the Apollo 11 mission (NASA/JSC). Neil Armstrong, Buzz Aldrin and Michael Collins brought it to the White House on the occasion of the 30th anniversary of the Moon landing. At President Clinton's request, fragment 10057 remained on display in the Oval Office until he left office in January 2001. A thin slice of 10057 is also embedded in the center of the Space Window stained glass mosaic at the National Cathedral in Washington, DC.

The Apollo 11 Lunar Module *Eagle* reached the Moon on July 20th, 1969, landing on the Sea of Tranquility. In the course of the 2.25 hours astronauts Neil Armstrong and Buzz Aldrin spent outside of the spacecraft, they collected 21.55 kg of soil and rock samples.

One of the largest rocks Apollo 11 astronauts brought back to Earth, labeled as basalt rock 10057, weighed nearly 1 kg and it was later cut into many pieces. The following table summarizes results of strontium and rubidium isotope measurements made by Papanastassiou et al. [1970, Table 1] on the basalt sample 10057 from the Sea of Tranquility. The texture of this rock indicates that its parent magma was extruded onto the lunar surface and cooled rapidly.[69]

[69] O. B. James and E. D. Jackson. Petrology of the Apollo 11 ilmenite basalts. *Journal of Geophysical Research*, 75(29):5793–5824, 1970. doi:10.1029/JB075i029p05793

ROCK PART	^{87}Rb/^{86}Sr	^{87}Sr/^{86}Sr
plagioclase	0.01567 ± 0.00024	0.70020 ± 0.00006
whole rock	0.10250 ± 0.00154	0.70469 ± 0.00003
ilmenite	0.20960 ± 0.00314	0.71023 ± 0.00004

Papanastassiou et al. [1970] report that the measured values of $N(^{87}\mathrm{Rb})/N(^{86}\mathrm{Sr})$ for the samples have "maximum errors of $\pm 1.5\,\%$." Since the same authors report expanded uncertainties (2σ) for the corresponding measured values of $N(^{87}\mathrm{Sr})/N(^{86}\mathrm{Sr})$, here we interpret the 1.5 % as relative expanded uncertainty for 95 % coverage, hence take the relative standard uncertainty to be 0.75 %. The symbol $N(^{87}\mathrm{Rb})$ denotes the number of atoms of ^{87}Rb in the aliquot of the material that was analyzed, and similarly for the other isotopes. For brevity, we will sometimes omit the quantity symbol that denotes "number of atoms", as we do in the table above.

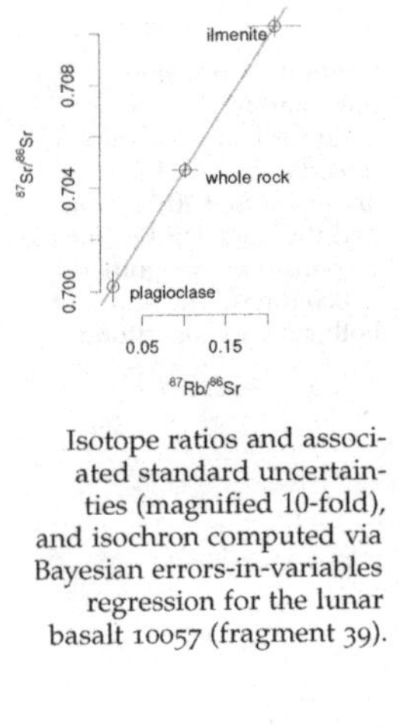

Isotope ratios and associated standard uncertainties (magnified 10-fold), and isochron computed via Bayesian errors-in-variables regression for the lunar basalt 10057 (fragment 39).

^{87}Rb is a naturally occurring, radioactive isotope of rubidium which decays to stable ^{87}Sr by emission of an electron and antineutrino (β^- decay).

As the magma cooled, the crystals that form will have

different levels of rubidium whereas the isotopic composition of strontium remains uniform.[70] Inspection of the entries under ^{87}Rb/^{86}Sr in the preceding table reveals that this ratio is quite variable for different aliquots of the same sample, varying by an order of magnitude from 0.016 to 0.210.

[70] P. H. Warren and G. J. Taylor. The Moon. In H. D. Holland and K. K. Turekian, editors, *Treatise on Geochemistry*, volume 2, pages 213–250. Elsevier, Oxford, UK, second edition, 2014

The rocks appear unaltered since their formation, except for superficial patina, micro-cratering, and exposure to cosmic radiation that have induced nuclear reactions in the samples.[71] Therefore, it seems safe to assume that, after consolidation of the magma into these basalts, they will have neither lost nor gained any of the isotopes of strontium or rubidium used for dating, other than for the changes attributable to the decay of ^{87}Rb.

[71] C. Meyer. The Lunar Sample Compendium. `https://curator.jsc.nasa.gov/lunar/lsc/`, 2011. NASA Astromaterials Research & Exploration Science

If there were $N_0(^{87}\mathrm{Sr})$ atoms in the magma originally, at any subsequent epoch t the following relationship should hold:

$$N_t(^{87}\mathrm{Sr}) = N_0(^{87}\mathrm{Sr}) + (e^{\lambda t} - 1) N_t(^{87}\mathrm{Rb}),$$

where $\lambda = \ln(2)/t_{½}(^{87}\mathrm{Rb})$ is the decay constant of ^{87}Rb. Since isotope ratios are usually easier to determine than the concentrations themselves, the more practically relevant relationship is this:

$$\frac{N_t(^{87}\mathrm{Sr})}{N_t(^{86}\mathrm{Sr})} = \frac{N_0(^{87}\mathrm{Sr})}{N_0(^{86}\mathrm{Sr})} + (e^{\lambda t} - 1)\frac{N_t(^{87}\mathrm{Rb})}{N_t(^{86}\mathrm{Sr})}.$$

In this equation, we consider $N_t(^{86}\mathrm{Sr})$ a constant, since ^{86}Sr is not radiogenic. Thus, $N_t(^{86}\mathrm{Sr}) = N_0(^{86}\mathrm{Sr})$.

Since components of the sample all are of approximately the same age, it follows that, once their values of ^{87}Sr/^{86}Sr are plotted (along the vertical axis) against corresponding values of ^{87}Rb/^{86}Sr (along the horizontal axis), they should be approximately aligned along a straight line, called an *isochron* (which means of "the same age").

The slope of this straight line is $\beta = e^{\lambda t} - 1$, which means that the crystallization age of the sample is

$$t = \ln(1+\beta)/\lambda.$$

Ordinary least squares is not the appropriate procedure to estimate the slope (and intercept) of this isochron because both isotope ratios are observed with error. We have already faced a similar challenge when comparing two measurement methods, which we addressed using Deming regression and Passing-Bablok regression.

Suppose that we wish to fit an isochron to n pairs of ratios (x_1, y_1), …, (x_n, y_n), where x_j denotes a value of $^{87}Rb/^{86}Sr$ and y_j denotes a value of $^{87}Sr/^{86}Sr$. We model these measured values of the isotope ratios as being equal to their corresponding true values plus measurement errors:

$$\begin{aligned} y_j &= \nu_j + \varepsilon_j, \\ x_j &= \xi_j + \delta_j, \text{ and} \\ \nu_j &= a + b\xi_j, \text{ for } j = 1, \ldots, n, \end{aligned}$$

where we assume that the measurement errors $\{\varepsilon_j\}$ and $\{\delta_j\}$ are non-observable outcomes of independent, Gaussian random variables with mean 0 and with standard deviations $\{u(y_j)\}$ and $\{u(x_j)\}$, respectively. We also assume that the true isochron, with intercept a and slope b, is a relationship between the true values of the ratios.

Unlike with ordinary least squares, note that this model has $2+n$ parameters: the intercept and slope of the regression line, and the true values, $\{\xi_j\}$, of the abscissas of the n data points. These parameters have to be estimated based on the n quadruplets $\{(x_j, u(x_j), y_j, u(y_j))\}$.

This model can be fitted to the data in any one of several different ways. For example, by the method of generalized distance regression, which involves nonlinear, numerical optimization of parameters a, b, and ξ by minimizing the following objective function:

$$\begin{aligned} S(a,b,\xi_1,\xi_2,\xi_3) = & \frac{(x_1-\xi_1)^2}{u^2(x_1)} + \frac{(y_1-(a+b\xi_1))^2}{u^2(y_1)} + \\ & \frac{(x_2-\xi_2)^2}{u^2(x_2)} + \frac{(y_2-(a+b\xi_2))^2}{u^2(y_2)} + \\ & \frac{(x_3-\xi_3)^2}{u^2(x_3)} + \frac{(y_3-(a+b\xi_3))^2}{u^2(y_3)}. \end{aligned}$$

A Bayesian (Page 204) formulation is generally preferable, if for no other reason because it regularizes the problem,[72] and effectively replaces optimization with a gentle, guided exploration of the set of possible values for the parameters. Below is the Bayesian formulation of the model in the Stan language.[73]

[72] T. Hastie, R. Tibshirani, and J. Friedman. *The Elements of Statistical Learning: Data Mining, Inference, and Prediction*. Springer-Verlag, New York, second edition, 2009. URL `statweb.stanford.edu/~tibs/ElemStatLearn/`

[73] Stan Development Team. *Stan Modeling Language — User's Guide and Reference Manual*. Available at `http://mc-stan.org/`, 2016. Stan Version 2.14.0

```
data {
  int<lower=0> n;        // Number of pairs of isotope ratios
  vector[n] x;           // Values of 87Rb/86Sr
  vector[n] ux;          // Standard uncertainties
  vector[n] y;           // Values of 87Sr/86Sr
  vector[n] uy;          // Standard uncertainties
  real aPriorMean;       // Prior mean of intercept
  real aPriorSD;         // Prior std. dev. of intercept
  real bPriorMean;       // Prior mean of slope
  real bPriorSD;         // Prior std. dev. of slope
  // Prior means and standard deviations for true values of {x}
  vector[n] xiPriorMean;
  vector[n] xiPriorSD;
}
parameters {
  real a;                // Intercept of isochron
  real b;                // Slope of isochron
  vector[n] xi;          // True values of {x}
}
model {
  // Prior distributions for isochron's intercept and slope
  a ~ normal(aPriorMean, aPriorSD);
  b ~ normal(bPriorMean, bPriorSD);
  // Prior for true isotope ratios 87Rb/86Sr
  xi ~ normal(xiPriorMean, xiPriorSD);
  // Likelihood for isotope ratios 87Rb/86Sr
  x ~ normal(xi, ux);
  // Likelihood for isotope ratios 87Sr/86Sr
  y ~ normal(a + b * xi, uy);
}
```

The following R codes assume that this Stan model has been assigned to variable `eivModel` as a character string, including the line breaks. We shall use the regression coefficients produced by the generalized distance regression, and their uncertainties, as soft guides for the Bayesian method. Thus, in our Bayesian model, the prior distributions reflect the belief that the slope and intercept should not to be too far from the generalized distance regression estimates, and that the true values of the ratios, $\{\xi_j\}$, also should not be too far from the observed values.

```
## Measurement results for lunar basalt 10057,39
z = data.frame(
        sample=c('plagioclase','whole rock','ilmenite'),
        x=c(0.01567, 0.10250, 0.20960), # 87Rb/86Sr
        ux=c(0.00024, 0.00154, 0.00314),
        y=c(0.70020, 0.70469, 0.71023), # 87Sr/86Sr
        uy=c(0.00006, 0.00003, 0.00004)
        )

## Generalized distance regression estimates
gdr = array(NA,dim=c(1e5, 2))
for (i in 1:1e5) {
    if (i %% 1000 == 0){cat(i, "of", 1e5, "\n")}
x.mc = rnorm(3, z$x, z$ux)
y.mc = rnorm(3, z$y, z$uy)
gdr[i,] = optim(par = c(0.7, 0.05, z$x),
  function(p) {
    xTrue = p[-c(1,2)]
    yTrue = p[1] + p[2]*xTrue
    sum( ((y.mc - yTrue)/z$uy)^2 + ((x.mc - xTrue)/z$ux)^2 )
    })$par[1:2]
}
gdr.coef = apply(gdr, 2, mean)
gdr.unc = apply(gdr, 2, sd)
```

Now we use the parameter estimates from the generalized distance regression as the initial values for the Bayesian method and to set the prior distributions.

```
require(rstan)
eiv.init = function () list(a=gdr.coef[1],
                            b=gdr.coef[2], xi=z$x)
eiv.data = data=list(n=3, x=z$x, ux=z$ux, y=z$y, uy=z$uy,
           aPriorMean=gdr.coef[1], aPriorSD=3*gdr.unc[1],
           bPriorMean=gdr.coef[2], bPriorSD=3*gdr.unc[2],
           xiPriorMean=z$x, xiPriorSD=3*z$ux)
eiv.stan = stan(model_code = eivModel,
                data = eiv.data,
                init=eiv.init, iter=50000)
print(eiv.stan, digits = 5)
```

The regression slope estimates from the two methods are nearly identical in our case, giving $b = 0.0517(8)$ for the generalized distance regression and $b = 0.0517(7)$ for the Bayesian method.

The basalt age estimate from the Sea of Tranquility and its associated uncertainty is obtained by using the draws from the posterior distribution of the isochron slope which are combined with the draws from a Gaussian distribution representing the half-live of ^{87}Rb with mean 49.7 Ga and standard deviation 0.3 Ga, which are the estimate and associated standard uncertainty from NUBASE2000 [Kondev et al., 2021].

```
eiv.stan.mcmc = rstan::extract(eiv.stan)
N.mcmc = length(eiv.stan.mcmc$b)

tRb = rnorm(N.mcmc, 49.7, 0.3)
lambda = log(2)/tRb
age = log(1 + eiv.stan.mcmc$b)/lambda
c(age=mean(age), age.u=sd(age),
  quantile(age, probs=c(0.025, 0.975)))
```

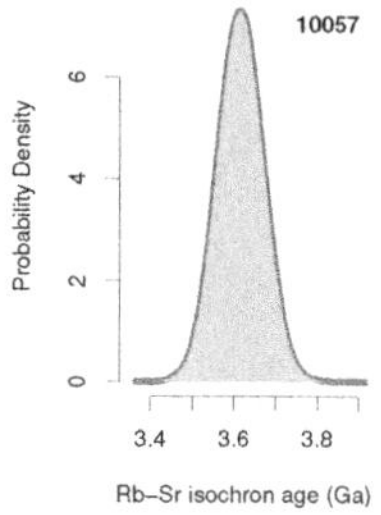

Posterior probability density (Page 159) of the Rb-Sr isochron age determined for the lunar basalt 10057 (fragment 39).

The resulting Rb-Sr isochron age of the lunar basalt 10057 is $t = 3.62$ Ga with the associated 95 % coverage interval 3.51 Ga to 3.72 Ga.

Prior to the lunar landings, there were doubts about whether there had ever been volcanic activity on the Moon but the discovery of these basalts, which are consolidated lava resulting from volcanic eruptions on the Moon more than 3 billion years ago, dispelled such doubts.

Paintings of Vermeer

In 1937, Abraham Bredius, one of the premier art historians of his time, announced *The Supper at Emmaus* — a new Vermeer painting that had recently been discovered in France:

Han van Meegeren (1937) *The Supper at Emmaus.* Museum Boijmans van Beuningen, Rotterdam (Wikimedia Commons).

> It is a wonderful moment in the life of a lover of art when he finds himself suddenly confronted with a hitherto unknown painting by a great master [...] Quite different from all his other paintings and yet every inch a Vermeer.[74]

Many leading art-historians soon accepted this "early" Vermeer, and the painting was purchased by the Rembrandt Society as a gift to the Museum Boijmans in Rotterdam, for an amount comparable to 5 million of today's Euro.

[74] A. Bredius. A new Vermeer. *The Burlington Magazine for Connoisseurs*, 71(416): 210–211, November 1937

Shortly after the end of World War II, investigators tracing the provenance of alleged Vermeer paintings held by the Nazi Field-Marshall Hermann Goering were led to Han van Meegeren. They uncovered multiple "masterpieces" that were, in fact, elaborate forgeries, among them *The Supper at Emmaus.*

Eventually van Meegeren was tried and convicted of forgery and fraud in 1947, but doubts lingered as to the possible authenticity of some of the questionable paintings. The matter was settled only much later, using radioisotopes that enabled estimating the date when the lead was extracted from the ore for use as a white pigment in the paint.[75]

[75] B. Keisch. Dating works of art through their natural radioactivity: Improvements and applications. *Science*, 160(3826):413–415, 1968. doi:10.1126/science.160.3826.413

This method for distinguishing between modern and old artworks is based on measuring the radioactivity of radium-226 and polonium-210 (which acts as a proxy for the hard-to-measure lead-210) in lead pigment samples.

Natural lead contains the radioactive lead-210 isotope which is constantly supplied from the radioactive decay of radium-226. For this reason, old samples of lead pigments will show equal amounts of radioactivity from lead-210 and radium-226. During the smelting of lead ores, most of the radium is removed instantly and it takes more than a century for the excess lead-210 to decay. Hence, modern samples will have much larger amounts of radioactivity from lead-210 compared to radium.

The half-life of lead-210, $t_{½}(^{210}\mathrm{Pb})$, is 22.20 years with standard uncertainty 0.22 years[Audi et al., 2017].

The half-life of radium-226, $t_{½}(^{226}\mathrm{Ra})$, is 1600 years with standard uncertainty 7 years[Audi et al., 2017].

The ratio of the specific activities of these two isotopes is a function of the age (t) of the lead used in the paint and derives from the underlying physical model of radioactive decay that follows consecutive radioactive decay reactions:

The radioactive decay rate constants are related to the nuclide half-life. For example, $t_{½}(^{210}\mathrm{Pb}) = \ln(2)/k_{\mathrm{Pb}}$.

$$^{226}\mathrm{Ra} \xrightarrow{k_{\mathrm{Ra}}} {}^{210}\mathrm{Po} \quad \text{or} \quad {}^{210}\mathrm{Pb} \xrightarrow{k_{\mathrm{Pb}}} {}^{210}\mathrm{Bi}.$$

Solving the differential equations that correspond to this physical model gives the following ratio of $^{210}\mathrm{Po}$ and $^{226}\mathrm{Ra}$ at any given time:

$$\frac{A(^{210}\mathrm{Po})}{A(^{226}\mathrm{Ra})} \approx K + (F - K)\exp\{(k_{\mathrm{Ra}} - k_{\mathrm{Pb}})t\},$$

where $K = k_{\mathrm{Pb}}/(k_{\mathrm{Pb}} - k_{\mathrm{Ra}}) \approx 1$ and F is the estimate of the value that the ratio $A(^{210}\mathrm{Po})/A(^{226}\mathrm{Ra})$ would have had when the lead was smelted. The value of this *separation factor* needs to be determined in order to calibrate the measurement model.

Keisch et al. [1967] list data used to determine the separation factors (F) from materials including lead from an old English pipe and white lead pigments of known manufacture dates. Some of these observations are shown in the table below.

Sample	Year (D)	$A(^{210}\mathrm{Po})$	$A(^{226}\mathrm{Ra})$
Lead pipe (England)	16th c.	0.039 ± 0.041	0.08 ± 0.03
Portrait (Italy)	1600	0.21 ± 0.10	0.21 ± 0.29
Female saint (Italy?)	1750–1800	3.8 ± 0.7	3.0 ± 0.4
Portrait (France)	1830–1840	5.3 ± 2.3	3.9 ± 1.4
Portrait (USA)	1921	6.3 ± 1.3	0.0 ± 0.3
Lead chloride (USA)	1962	46 ± 4	0.5 ± 0.4

Results of the analyses of white lead pigment and related lead-bearing materials with known dates of manufacture (smelting).[Keisch et al., 1967]

The values of $A(^{210}\mathrm{Po})$ and $A(^{226}\mathrm{Ra})$ are numbers of disintegrations per minute and per gram of lead, analyzed ca. 1966.

The separation factor is then calculated for all specimens of known age (which is 1966 − D years):

$$F = \frac{A_0(^{210}\text{Po})}{A_0(^{226}\text{Ra})},$$

where the estimated activities right after the smelting are as follows:

$$A_0(^{210}\text{Po}) = A(^{210}\text{Po}) \exp\{(D_0 - D)k_{\text{Pb}}\},$$
$$A_0(^{226}\text{Ra}) = A(^{226}\text{Ra}) \exp\{(D_0 - D)k_{\text{Ra}}\}.$$

To estimate the separation factor, F, and to evaluate its uncertainty, $u(F)$, for the 1921 portrait listed in the preceding table, we repeat the following steps 10^6 times:

(1) Draw a value for D_0 from a uniform distribution between 1966 and 1967.

(2) Draw a value for D from a uniform distribution between 1921 and 1922.

(3) Draw a value for $t_{1/2}(^{226}\text{Ra})$ from a Gaussian distribution with mean 1600 and standard deviation 7 truncated at zero, and calculate the corresponding rate constant value.

(4) Draw a value for $t_{1/2}(^{210}\text{Pb})$ from a Gaussian distribution with mean 22.20 and standard deviation 0.22 truncated at zero, and calculate the corresponding decay rate constant value.

(5) Draw a value for $A(^{226}\text{Ra})$ from a Gaussian distribution with mean 0.0 and standard deviation 0.3 truncated at zero.

(6) Draw a value for $A(^{210}\text{Po})$ from a Gaussian distribution with mean 6.3 and standard deviation 1.3 truncated at the value of $A(^{226}\text{Ra})$ from step (5).

(7) Compute $\ln(F)$ using the results from steps (1)-(6).

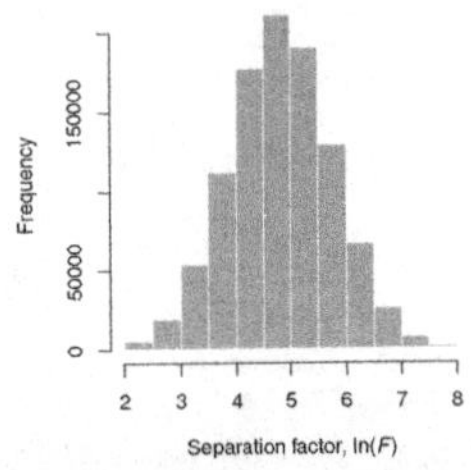

Histogram of 10^6 replicates of the value of $\ln(F)$ estimated from the lead pigment in the 1921 portrait (sample M-17-H) [Keisch et al., 1967].

Then we compute the median, m, of the 10^6 values of $\ln(F)$, and the rescaled median of the absolute deviations from the median (mad) of the $\ln(F)$ values, s.

The procedure just described for one particular sample of paint is then repeated for all the available 71 materials that Keisch et al. [1967] list in Tables 2-4. The resulting 71 Gaussian distributions of $\ln(F)$ were combined using the linear pool [Koepke et al., 2017], with the result shown alongside, which can be approximated with a Gaussian distribution whose mean and standard deviation are the median $m = 4.8$ and rescaled median of the absolute deviations from the median (mad) $s = 1.7$.

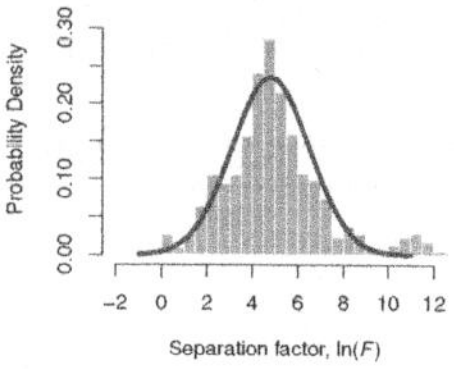

The distribution of experimentally determined separation factors, $\ln(F)$, with a fitted Gaussian distribution.

The model age of paintings can therefore be estimated by solving the above measurement model for t, by entering measured values of the activities of the two isotopes:

$$t = \frac{1}{k_{\mathrm{Pb}} - k_{\mathrm{Ra}}} \left\{ \ln(F - K) - \ln\left(\frac{A(^{210}\mathrm{Po})}{A(^{226}\mathrm{Ra})} - K \right) \right\}.$$

To simplify the model, we do not take into account the radioactive decay of atoms of radium, which is negligible in the time scale considered here, hence set $k_{\mathrm{Ra}} = 0$ and $K = 1$ without any noticeable loss of accuracy.

The age estimates and their uncertainties are then obtained using the Monte Carlo method in very much the same way as it was used for the separation factor.

PAINTER	WORK	$A(^{210}\mathrm{Po})$	$A(^{226}\mathrm{Ra})$	AGE / YEARS
van Meegeren	WF	12.6 ± 0.7	0.26 ± 0.07	51 ± 41
van Meegeren	SE	8.5 ± 1.4	0.8 ± 0.3	87 ± 52
Vermeer	LM	1.5 ± 0.3	1.4 ± 0.2	212 ± 65
Vermeer	LG	5.2 ± 0.8	6.0 ± 0.9	232 ± 67

WF *The Washing of the Feet*, Rijksmuseum, Amsterdam
SE *The Supper at Emmaus*, Museum Boijmans, Rotterdam
LM *The Lacemaker*, Musée du Louvre, Paris
LG *Officer and Laughing Girl*, The Frick Collection, New York

Measurements of specific activity of white lead pigments [Keisch, 1968], expressed in numbers of disintegrations per minute and per gram of lead, and corresponding mean ages and associated standard uncertainties from the Monte Carlo uncertainty evaluation.

The actual ages that the paintings had in 1968, when the study by Keisch [1968] was published, are well-documented – WF: 37 years, SE: 32 years, LM: 298 years, and LG: 311 years. These ages are generally consistent with the mean radioisotopic ages when the associated uncertainties are taken into account.

The following coverage intervals were derived from the samples produced by the Monte Carlo method, and include the true age of the lead pigment with 95 % confidence. In the image below, the open circles indicate the median radioisotopic age estimates of the four paintings. Both *The Washing of the Feet* and *The Supper at Emmaus* clearly contain lead pigment from the 20th century.

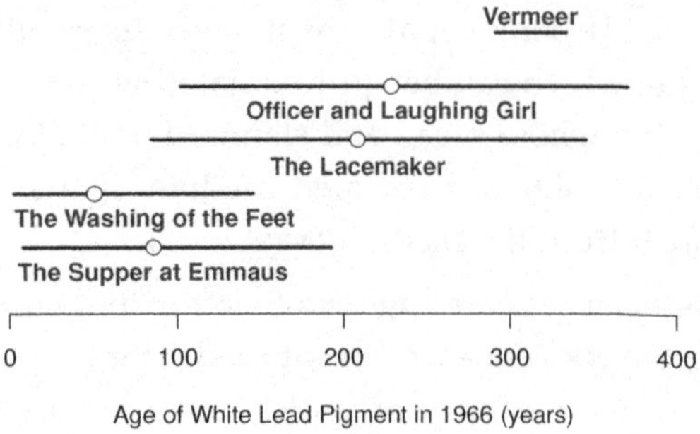

The line segment labeled Vermeer represents Vermeer's lifetime.

The results can also be summarized by estimating the *largest* model age for each of these paintings, such as their 99th percentiles. Both authentic paintings produce values consistent with Vermeer's lifetime (1632-1675) whereas both forgeries fall short by more than a century.

Shroud of Turin

Caravaggio (1603-1604) *La Deposizione di Cristo*, Pinacoteca Vaticana, Vatican City — Wikimedia Commons

The discovery of *radiocarbon dating* earned Willard F. Libby the 1960 Nobel Prize in Chemistry, and the accolade from the Nobel Committee that "seldom has a single discovery in chemistry had such an impact on the thinking in so many fields of human endeavor."

^{14}C atoms are continuously generated in Earth's atmosphere as neutrons produced by cosmic rays strike nitrogen atoms, and eventually are absorbed by living organisms. The concentration of ^{14}C in the living tissues stays in equilibrium with its atmospheric counterpart until the organism dies. Thereafter, the ratio of concentrations of ^{14}C and of ^{12}C in the remains decreases steadily over time.

By measuring this ratio in the remains, and assuming that the ratio of concentrations of ^{14}C and ^{12}C in the atmosphere during the organism's lifetime was the same as it is today, it is possible to estimate how long ago the plant or animal died.

While simple in principle, radiocarbon dating is challenging in practice. First, the amount fraction of ^{14}C in pure carbon is minuscule: about 1 atom of ^{14}C per trillion atoms of carbon (of which the vast majority are ^{12}C and ^{13}C atoms). This implies that, in 4 grams of carbon, only one atom of ^{14}C will decay per second, on average. Therefore, radiocarbon dating based on measurements of activity requires fairly large samples of material. Mass spectrometry, which actually counts atoms of different mass numbers, has enabled radiocarbon dating of very small samples of material.

Second, radiocarbon dating rests on two key assumptions: (i) that the ratio of concentrations of ^{14}C and ^{12}C atoms in the atmosphere has remained constant over time, and equal to its present value; and (ii) that its value is the same for all biological tissues. Neither of these assumptions is valid. The first because the burning of fossil fuels (which contain no ^{14}C) has steadily decreased the fraction of ^{14}C in the atmosphere, while detonations of nuclear weapons from the 1940s until the early 1960s, increased it. The second because isotopic fractionation changes the relative concentrations of the three isotopes of carbon according to the provenance of the biological material used for dating.

Positive and negative versions of a portion of the Shroud of Turin — WikiMedia Commons

These contingencies imply that accurate dating cannot be achieved without calibration, which establishes a correspondence between radiocarbon ages based on the ideal assumptions aforementioned, and known calendar ages of particular samples.

The most recent calibration curve is INTCAL2020.[76] For the most recent 14 000 years, this curve is based entirely on tree-ring measurements, which can be dated by counting rings from outermost to innermost. Also, each ring's isotopic composition is a snapshot of the atmospheric

[76] T. J. Heaton, M. Blaauw, P. G. Blackwell, C. Bronk Ramsey, P. J. Reimer, and E. M. Scott. The INTCAL20 approach to radiocarbon calibration curve construction: a new methodology using Bayesian splines and errors-in-variables. *Radiocarbon*, pages 1–43, 2020. doi:10.1017/RDC.2020.46

composition at the time when the ring was growing.

The measurement of the age of the Shroud of Turin using radiocarbon dating is one of the most talked-about applications of the technique. The shroud is a linen cloth kept in the Cathedral of Saint John the Baptist, in Turin, Italy, which bears marks of the body of a tall, bearded man who may have been flogged. Some people believe that it is the burial cloth of Jesus of Nazareth.

By convention, radiocarbon ages are expressed as numbers of years before 1950, denoted as "before present" (BP) although there is some uncertainty about whether this means the the very beginning of 1950 [Townsley, 2017] or mid-1950 [Ramsey, 2009].

Mass spectrometric measurements made in 1988 by Damon et al. [1989] at laboratories in Tucson (Arizona, USA), Oxford (UK), and Zurich (Switzerland), yielded average radiocarbon age of 691 years Before Present (BP), with standard uncertainty 31 years.

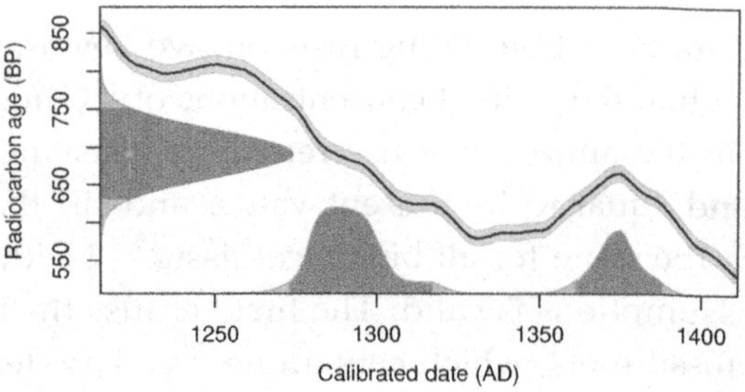

Calibration of the 1988 measurement of the radiocarbon age of the Shroud of Turin using the INTCAL2020 [Reimer et al., 2020] calibration curve, to obtain an estimate of the calendar age.

The radiocarbon age was "translated" into calendar age using R function `calibrate` defined in package `clam`.[77] The resulting distribution of calendar age is a bizarre bimodal distribution whose mean (AD 1317) and standard deviation (40 years) tell us very little about the likely age of the shroud, providing a cogent illustration of the fact that probability densities are well suited to capture the uncertainty of complex outcomes whereas summary estimates can be spectacularly deceiving.

77 M. Blaauw. Methods and code for 'classical' age-modelling of radiocarbon sequences. *Quaternary Geochronology*, 5:512–518, 2010. doi:10.1016/j.quageo.2010.01.002; and M. Blaauw. *clam: Classical Age-Depth Modelling of Cores from Deposits*, 2021. URL `https://CRAN.R-project.org/package=clam`. R package version 2.4.0

The age provided by the R function `calibrate` lies between AD 1273 and AD 1317 with 65 % probability, and between AD 1361 and AD 1388 with 30 % probability – hence the shroud is Medieval and not from Antiquity.

Categorizing

Nominal and *ordinal* properties are kinds of *categorical* properties, which are qualitative [Agresti, 2019].

The identity of a polychlorinated biphenyl (PCB), and the species of a maple tree (genus *Acer*), are nominal properties. The former has more than 200 possible values, the latter more than 160. The only meaningful comparison between values of the same nominal property is whether they are identical or different.

The values of a nominal property are names of sets of entities that have the same values of the attributes that characterize the nominal property.

For example, when presented with an animal of the genus *Panthera*, one compares it with standard specimens of the five species in this genus, to determine whether the animal is a tiger, leopard, jaguar, lion, or snow leopard.

This comparison may involve examining qualitative attributes such as the body shape, the color of the fur, or the footprint. It may also involve examining quantitative attributes, like body length, height, or mass. If only a sample of tissue from the animal is available, then the comparison may involve sequencing particular areas of the genome, and comparing these sequences with paradigmatic sequences of known provenance that are available in gene databases.[78]

[78] Y. Cho, L. Hu, H. Hou, et al. The tiger genome and comparative analysis with lion and snow leopard genomes. *Nature Communications*, 4:2433, September 2013. doi:10.1038/ncomms3433

Ordinal properties have values that can be ordered (that is, ranked) from smallest to largest, or from lowest to highest, but for which neither differences nor ratios are meaningful, even when their values are represented numerically.

Cancer stage, as defined by the American Joint Committee on Cancer, is a property of breast cancer, whose possible values are the Roman numerals I, II, III, and IV. However, one is not entitled to say either that stage IV is two times "worse" than stage II, or that the difference in severity between stages III and I is the same as the difference in severity between stages IV and II.

The Mohs hardness of a mineral is expressed relative to a scale ranging from 1 (for talc) to 10 (for diamond): however, neither is fluorite (4) two times harder than gypsum (2), nor is the difference in hardness between topaz (8) and apatite (5) the same as the difference in hardness between quartz (7) and fluorite (4).

Mohs hardness can also be expressed using a number half-way between any two consecutive integers in that range. For example, since tourmaline typically scratches quartz (Mohs hardness 7) and it is scratched by topaz (Mohs hardness 8), its Mohs hardness is conventionally designated as 7.5. However, this is only a way of saying that its hardness lies between the hardness of quartz and the hardness of topaz.

Measuring Abortion Rates

Unsafe abortion caused 5 % to 13 % of maternal deaths worldwide during 2010–2014, and a large proportion of the abortions were performed unsafely.[79] The prevalence of abortion therefore is an important public health measurand. Having ever had an induced abortion is a nominal property of every woman, whose values are YES or NO. Determining its value reliably is challenging because women often are reluctant to report it.

[79] WHO. Preventing unsafe abortion. Evidence Brief WHO/RHR/19.21, World Health Organization, Geneva, Switzerland, 2019

In a randomized response, house-to-house survey conducted in Mexico City in 2001, each participating woman was asked one of two questions, selected at random, as if by tossing a fair coin: whether she had ever attempted to terminate a pregnancy, or whether she was born in the month of April.[80]

[80] D. Lara, J. Strickler, C. D. Olavarrieta, and C. Ellertson. Measuring induced abortion in Mexico: A comparison of four methodologies. *Sociological Methods & Research*, 32 (4):529–558, May 2004. doi:10.1177/0049124103262685

Only the woman being interviewed could see which of these two questions had been drawn for her, and she truthfully answered YES or NO to the question she was presented with. Since this survey technique preserves confidentiality, it tends to produce more reliable results than, for example, interviews where a woman is asked directly, face-to-face, the sensitive question about abortion.

Of the 250 women that participated in the house-to-house survey, 33 answered YES to the question they were presented with. This number includes women who had had an abortion and were asked the question about abortion, as well as women who were born in the month of April and were asked whether it was so, regardless of whether they had ever had an abortion. Since the survey design prevents determining individual values of the nominal property, the goal is to measure its prevalence, α, which is the proportion of women who had ever attempted an abortion.

This type of survey safeguards the confidentiality of responses and by doing so improves the reliability of its results. However, confidentiality could possibly be breached if the interviewer knew the participant personally, and also knew that she was not born in April. In such case, a YES answer reveals the attempted abortion.

The following diagram shows how YES and NO answers may arise, where $p = 1/2$ is the probability of being asked the sensitive question, and $q = 1/12$ denotes the probability of having been born in April. The last column lists the probabilities of the different instances of YES and NO. Note that the six probabilities sum to 1.

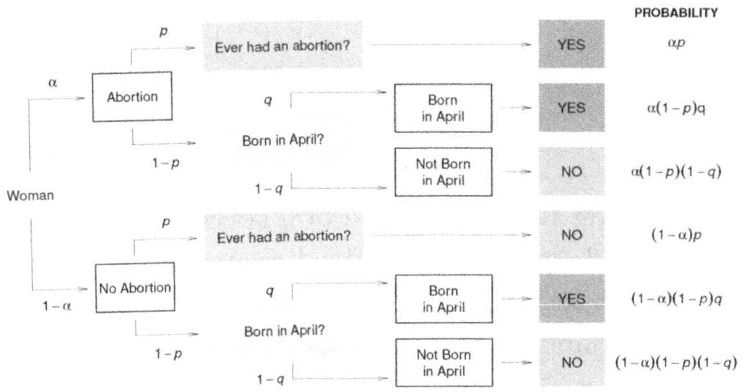

Randomized response survey to measure prevalence of abortion.

The probability of YES is θ, which is the sum of the three terms above that appear next to the rectangles shaded dark gray in the last column of the diagram:

$$\theta = \alpha p + \alpha(1-p)q + (1-\alpha)(1-p)q = \alpha p + q(1-p).$$

Since the estimate of θ is $\widehat{\theta} = 33/250$, $p = 1/2$ by design, and $q = 1/12$ on the assumption that births are equally likely to fall on any month of the year, α can be estimated by solving $\widehat{\theta} = \alpha p + q(1-p)$ for α, which yields

$$\widehat{\alpha} = \widehat{\theta}/p - q(1-p)/p = 271/1500 = 0.18.$$

The uncertainty associated with $\widehat{\alpha}$ is the same as the uncertainty associated with $\widehat{\theta}/p$ because both p and q are known with full certainty, hence so is $q(1-p)/p$. For the same reason, $u(\widehat{\theta}/p) = u(\widehat{\theta})/p$.

Now, the random variable $\widehat{\theta}$ has a binomial distribution (Page 173) based on 250 trials, whose variance can be estimated as $\widehat{\theta}(1-\widehat{\theta})/250$, with $\widehat{\theta} = 33/250 = 0.132$. Therefore,

$$u(\widehat{\alpha}) = u(\widehat{\theta})/p = \sqrt{\frac{0.132(1-0.132)}{250}}/(1/2) = 0.043.$$

A 95 % coverage interval for α can be derived from a corresponding coverage interval for θ, which can be computed as described under *Counts* (Page 177), finally to obtain $(0.10, 0.28)$, which is the output of the following R command:

```
theta = prop.test(x=33, n=250)$conf.int
p = 1/2; q = 1/12; (theta - q*(1-p))/p
```

Considering that each value of q specifies one particular model for the randomized response survey, the uncertainty in q may be incorporated via *model-averaging* (Page 139), and using the statistical bootstrap (Page 179). Assuming that q has a uniform distribution (Page 167) between $1/12.9$ and $1/12.6$ (the extreme birth rates for Mexico in the month of April during the period 2011-2017), the estimate of the prevalence of abortion becomes $\widetilde{\alpha} = 0.19$, and a 95 % uncertainty interval for the true value of α now ranges from 0.11 to 0.27.

The above estimate appears to stand in sharp contrast with the United Nations estimate of 0.1 per 1000 for the overall abortion rate in Mexico in 2003 (`data.un.org`, accessed April 7th, 2021), and with other estimates. A study of the incidence of abortion in Mexico conducted by El Colegio de Mexico, the Guttmacher Institute, and the Population Council Mexico Office,[81] reported that the overall abortion rate in Mexico, for 2006, was 33 per 1000 women aged 15-44, and a follow-up study[82] reported that the abortion rate in the region of Mexico City, for 2009, was 54 per 1000 women in the same age group.

[81] F. Juarez, S. Singh, S. G. Garcia, and C. D. Olavarrieta. Estimates of induced abortion in Mexico: what's changed between 1990 and 2006? *International Family Planning Perspectives*, 34:158–168, 2008. doi:10.1363/ifpp.34.158.08

All of the latter estimates are for incidence, supposedly per year, while the survey that Lara et al. [2004] conducted asked whether women had *ever* attempted an abortion, regardless of when, hence is more reflective of prevalence. If one divides the estimate of α above by the age range $44 - 15 = 29$ years, then one obtains a coarse proxy of 7 per 1000 for the incidence rate, which is intermediate between the United Nations estimate for 2003 and the more recent estimates quoted above.

[82] F. Juarez and S. Singh. Incidence of induced abortion by age and state, Mexico, 2009: new estimates using a modified methodology. *International Perspectives on Sexual and Reproductive Health*, 38:58–67, June 2012. doi:10.1363/3805812

Measuring Lethal Dose

The use of toxic chemicals is a normal part of contemporary life. They are used by farmers and doctors alike to eliminate pests in the field or eradicate diseases that stand in the way of our prosperity.

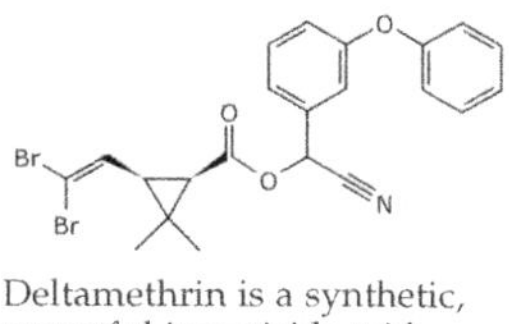

Deltamethrin is a synthetic, powerful insecticide with chemical structure similar to the natural pyrethrins produced by flowers such as the common daisy. While harmful to aquatic life, it is best known as the main agent in insecticide-treated mosquito nets used to fight malaria.

Evaluating the toxicity of a chemical typically involves exposing a model organism to various levels of the chemical and observing whether it survives. The outcome of such a study is a nominal property of the organism at each level of the toxin, whose values are DEAD or ALIVE, indicating the ultimate fate of the organism.

An experiment involving 5 batches of 100 freshwater mussels each was carried out to measure the toxicity of the insecticide deltamethrin, and yielded the following counts of dead and alive mussels after 24 h exposure to deltamethrin, at each of five concentration levels.[83]

[83] K. Köprücü and E. Seker. Acute toxicity of deltamethrin for freshwater mussel, *Unio elongatulus eucirrus* Bourguignat. *Bulletin of Environmental Contamination and Toxicology*, 80:1–4, 2008. doi:10.1007/s00128-007-9254-z

LEVEL OF DELTAMETHRIN	MUSSELS DEAD	MUSSELS ALIVE
5 µg/L	12	88
6 µg/L	19	81
7 µg/L	25	75
8 µg/L	37	63
16 µg/L	92	8

Assuming that the outcomes for different mussels in each batch of 100 are independent, and that the probability of death is the same for all the mussels in the same batch, then the number, y_i, of dead mussels in the batch exposed to concentration c_i has a binomial distribution (Page 173) based on 100 trials, with probability of death $p(c_i)$ for $i = 1, \dots, 5$.

The statistical measurement model is the following observation equation:

$$y_i = D_{i,1} + \cdots + D_{i,100}, \quad \text{for } i = 1, \dots, 5,$$

where $D_{i,j}$ is either 1 or 0, indicating whether mussel $j = 1, \dots, 100$ in batch i was dead or alive at the end of the experiment.

In many cases of exposure to potentially lethal agents, be they toxins or ionizing radiation, the probability of death first increases slowly with increasing dose, followed by a rapid increase, finally leveling off as the dose continues to increase, slowly approaching an upper bound that may be 100 %. Therefore, a plot of the probability of death against dose tends to be an S-shaped curve that may be fitted using several different functions.

The logistic regression model has a curve of this shape, and its parameters are easily interpretable. It is a *generalized linear model*[84] (Page 200) that expresses the log-odds (Page 161) of death as a linear function of the log dose:

[84] P. McCullagh and J. A. Nelder. *Generalized Linear Models*. Chapman & Hall / CRC, London, UK, 2nd edition, 1989

$$\ln\left\{\frac{p(c)}{1 - p(c)}\right\} = \alpha + \beta \ln(c),$$

where, in the present case, c denotes the mass concentration of deltamethrin, and $p(c)$ denotes the corresponding probability of death for the mussels exposed to it.

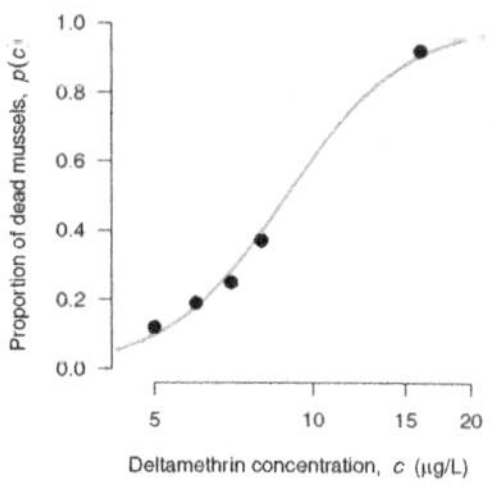

Mortality of freshwater mussels after 24 h exposure to various levels of deltamethrin. The S-shaped curve is the logistic regression curve with parameters $\alpha = -8.45$ (intercept) and $\beta = 3.85$ (slope).

The value of α determines the mortality corresponding to $c = 1\,\mu g/L$. In this case, the odds of death are $\exp(\alpha)$. For $\alpha = -8.45$, this is approximately 2 deaths per 10 000 mussels thus exposed. If the concentration of the insecticide is increased k-fold, then the odds of death increase k^β-fold. Given the estimate of $\beta = 3.85$, a 2-fold increase in concentration leads to a 14-fold increase in the odds of death.

Two popular alternatives to the logistic regression are the probit and the complementary log-log (CLOGLOG) models:

$$\Phi^{-1}(p) = \alpha + \beta \ln(c),$$
$$\ln\big(-\ln(1-p)\big) = \alpha + \beta \ln(c).$$

The function Φ^{-1} is the mathematical inverse of the cumulative distribution function of the Gaussian distribution with mean 0 and standard deviation 1. In R, values of $\Phi^{-1}(p)$ are obtained by executing qnorm(p).

McCullagh and Nelder [1989] point out that "the logistic and the probit function are almost linearly related over the interval" $0.1 \leqslant p \leqslant 0.9$, hence "it is usually difficult to discriminate between these two functions on the grounds of goodness-of-fit." These three models can be fit to the data as follows:

```
D = list(y=cbind(dead=c(12, 19, 25, 37, 92),
                 alive=c(88, 81, 75, 63, 8)),
         c=c(5, 6, 7, 8, 16) )

summary(glm(y ~ log(c), data=D, family=binomial(link=logit)))
summary(glm(y ~ log(c), data=D, family=binomial(link=probit)))
summary(glm(y ~ log(c), data=D, family=binomial(link=cloglog)))
```

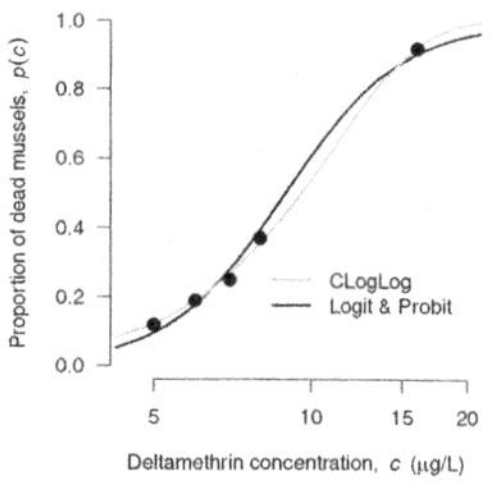

Fitting the three models to the mussel mortality data. The nearly overlapping lines are the logit and probit models. The gray line is the CLOGLOG model.

The values of the Bayesian Information Criterion (BIC, Page 100) — a model selection criterion whose smallest value indicates the best model — are 27.4 for the logit, 27.8 for the probit, and 25.9 for the CLOGLOG model. The slope, β, of the CLOGLOG model, is estimated as 0.1909. This implies that mortality increases $\exp(0.1909) = 1.21$ times per unit of increase of $\ln(c)$.

The toxic potency of a chemical can be expressed by a single numerical index, of which the most common is the concentration that is fatal to half of those exposed, known as the *median lethal dose* (LD_{50}). Other *toxic doses* can be defined similarly: for example, LD_{10} is the concentration that induces the death of 10 % of the exposed mussels, and LD_{90} is the concentration that induces the death of 90 % of the exposed mussels.

For the logistic model fitted above, the logarithm of LD_{10} can be computed as

$$\ln(\text{LD}_{10}) = \frac{\ln\left(\frac{0.1}{1-0.1}\right) - (-8.45)}{3.85} \approx 1.62.$$

R function `dose.p` from package `MASS`[85] computes the logarithms of the lethal doses for specified proportions of deaths, and evaluates the associated uncertainties:

[85] W. N. Venables and B. D. Ripley. *Modern Applied Statistics with S*. Springer, New York, Fourth edition, 2002. ISBN 0-387-95457-0. URL `www.stats.ox.ac.uk/pub/MASS4`

	LOGIT		PROBIT		CLOGLOG	
	$\ln(\text{LD}_p)$	u	$\ln(\text{LD}_p)$	u	$\ln(\text{LD}_p)$	u
$p=0.1$	1.62	0.05	1.63	0.04	1.53	0.06
$p=0.5$	2.19	0.03	2.20	0.03	2.26	0.03
$p=0.9$	2.76	0.08	2.76	0.07	2.73	0.05

Note that CLOGLOG yields the largest uncertainty for $\ln(\text{LD}_{10})$ and the smallest uncertainty for $\ln(\text{LD}_{90})$. The reason is the asymmetry of the CLOGLOG curve that relates probability of death to concentration of deltamethrin: it is the steepest of the three for high values of the concentration, and it is the flattest for low values of the concentration.

The physiological response, as summarized by LD_{50}, is also used as a measure of potency. Thus, it offers a way to measure the concentration of biomolecules, which is difficult to do using chemical methods. The standardization of the diphtheria-tetanus-pertussis vaccine is achieved this way.

Due to ethical concerns, toxicity assessments typically are based on small numbers of data points. This constraint, together with the nonlinear nature of the dose-response curves, make the interpretation of the results from such experiments dependent on model selection, a topic addressed under Model Uncertainty (Page 133).

Tinaroo virus

The above example of measuring the lethal dose can become more complex if the experiment captures more than just a binary outcome. Here we explore the toxicity of the tinaroo virus on chicken embryos for which three outcomes were recorded: *normal*, *deformed*, and *dead* after 18 days of exposure to the virus. The following observations, made in the course of a study conducted in Australia that looked at the family of arboviruses using embryonated chicken eggs as a model system for cattle and sheep.[86]

VIRUS	CHICKEN EMBRYOS		
LEVEL	NORMAL	DEFORMED	DEAD
3	18	0	1
30	17	0	2
2400	2	9	4
88 000	0	10	9

[86] D. A. McPhee, I. M. Parsonson, A. J. Della-Porta, and R. G. Jarrett. Teratogenicity of Australian Simbu serogroup and some other Bunyaviridae viruses: the embryonated chicken egg as a model. *Infection and Immunity*, 43:413–420, 1984. doi:10.1128/iai.43.1.413-420.1984

The level of the disease is an ordinal property whose values (*normal*, *deformed*, and *dead*) are determined by examination of the chicken embryos on day 18.

Suppose that the three ordered categories have probabilities — $p_1(E)$ for *normal*, $p_2(E)$ for *deformed*, and $p_3(E)$ for *dead* — that depend on the level (E) of exposure to the virus. Taken together, these probabilities define a probability distribution for the ordinal property Y, which is the severity of the infection.

Measuring this ordinal property means assigning a value to Y based on the observation of the corresponding exposure E. That is, Y is the the output quantity in a traditional measurement model, and E is an input quantity, except that here Y an ordinal quality and the model is a statistical measurement model [JCGM GUM-6:2020, §11].

The model specifies the probabilities of the different severity levels of the infection as function of the exposure levels, and, once calibrated, serves to assign the level of severity that has the highest probability, given the exposure level E. For the purposes of this example, we will assume that the virus levels are free from error, and that measurement uncertainty derives solely from the imperfect, stochastic relation between Y and E.

The probability that the severity of the infection on a particular chicken embryo (Y) will be less than or equal to j is

$$\gamma_j(E) = \Pr(Y \leqslant j) = p_1(E) + \cdots + p_j(E),$$

where $j = 1, 2$, and 3 denote *normal*, *deformed*, and *dead*, respectively. The statistical model we will fit to the data specifies that

$$\ln\left\{\frac{\gamma_j(E)}{1-\gamma_j(E)}\right\} = \theta_{j-1,j} - \beta\ln(E) \quad (j = 1,2,3),$$

where the ratio $\gamma_j(E)/(1-\gamma_j(E))$ is the odds (Page 161) of $Y \leqslant j$, and where θ are ordered *thresholds*:

$$(\theta_{0,1} = -\infty) \leqslant \theta_{1,2} \leqslant \theta_{2,3}.$$

This model is a *cumulative link model*[87] because it assumes that the values of the "link" function (in this case, the logit) of the cumulative probabilities of the ordered disease severity levels are determined by the values of the covariate E and the parameters $\theta_{1,2}$, $\theta_{2,3}$, β. The values of the covariate are observed, and the values of the parameters will have to be estimated from the data.

[87] A. Agresti. *Analysis of Ordinal Categorical Data*. John Wiley & Sons, Hoboken, NJ, 2nd edition, 2010. ISBN 978-0-470-08289-8. doi:10.1002/9780470594001

Suppose that E_1 and E_2 are two different values of the virus levels. The logarithm of the ratio of the corresponding odds of both suggesting the same level, j, of severity of the infection, is $\beta(E_2 - E_1)$, hence is independent of the severity level, and depends only on the difference between the exposure levels. For this reason, this model is called the *proportional odds model*[88].

[88] P. McCullagh. Regression models for ordinal data. *Journal of the Royal Statistical Society, Series B (Methodological)*, 42(2):109–142, 1980. doi:10.1111/j.2517-6161.1980.tb01109.x

We fit the model to the data above using R function `clm` defined in package `ordinal` [Christensen, 2019], by executing the following R code:

```
    virus = c(3, 20, 2400, 88000)
   normal = c(18, 17, 2, 0)
 deformed = c(0, 0, 9, 10)
     dead = c(1, 2, 4, 9)

exposure = c(rep(virus, normal),
             rep(virus, deformed), rep(virus, dead))
severity = c(rep(rep("normal",4), normal),
             rep(rep("deformed",4), deformed),
             rep(rep("dead",4), dead))

df = data.frame( logexposure=log(exposure),
                 severity=factor(severity, ordered=TRUE,
                 levels=c("normal", "deformed", "dead")) )

require(ordinal)
fit.clm = clm(severity ~ logexposure,
                         data=df, link="logit")

summary(fit.clm)
confint(fit.clm, type = "Wald", level=0.95)

## Confusion Matrix
table(df$severity, predict(fit.clm, type="class")$fit)
```

The resulting *confusion matrix* shows the number of observed disease severity in each row and the predicted severity in each column. For example, of the 37 embryos diagnosed as *normal*, the fitted model classified 35 correctly, but it misclassified 2 as *deformed*.

	NORMAL	DEFORMED	DEAD
NORMAL	35	2	0
DEFORMED	0	9	10
DEAD	3	4	9

The thresholds of this model are as follows:

$$\widehat{\theta}_{1,2} = 3.1 \pm 1.3 \text{ and } \widehat{\theta}_{2,3} = 5.4 \pm 1.7,$$

where each estimate is qualified with its 95 % confidence interval as calculated using the R code above.

The model's value assignments are based on the probabilities that the model computes for each severity level, given the value of the exposure. These probabilities are determined by the thresholds, $\theta_{1,2}$ and $\theta_{2,3}$, and by β, estimated as 0.51 with standard uncertainty 0.09.

The uncertainty associated with an assignment of severity level depends on how dispersed the unit of probability is over the three possible levels. This uncertainty can be quantified using the *entropy* of the corresponding probability distribution.

The entropy, H, of the probability distribution that puts probabilities $p_1(E)$, $p_2(E)$, and $p_3(E)$ to the *normal, deformed,* and *dead* levels, for a chicken embryo with exposure E to the tinaroo virus, is defined as

$$H = - \sum_{j=1,2,3} p_j(E) \ln p_j(E).$$

The larger the entropy, the greater the uncertainty in the assignment of severity level. Thus, entropy may be used to express the uncertainty associated with value assignments for ordinal properties.

The largest entropy possible in this case ($H = 1.1$) is achieved when $p_1(E) = p_2(E) = p_3(E) = 1/3$, and the lowest ($H = 0$) when one of these probabilities approaches 1 and the other two approach 0.

Consider the prediction this model makes for chicken embryos with exposure to $E = 10$ and 100 infective particles.

```
p = predict(fit.clm,
            newdata = data.frame(logexposure = log(c(10, 100))))
```

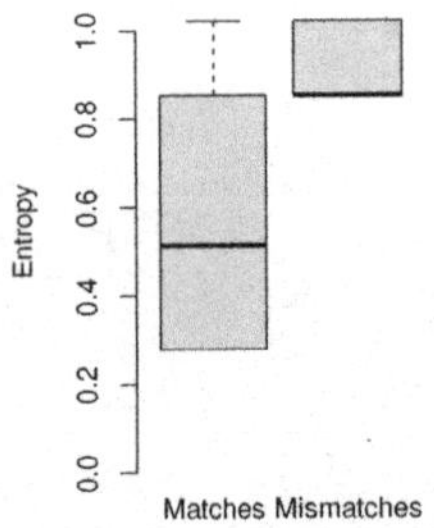

Entropy of the probability distributions produced by the cumulative link model as it classified cases as *normal, deformed,* or *dead.*

In both cases the predicted classification is *normal* but the uncertainties surrounding this prediction are very different:

	$p(E = 10)$	$p(E = 100)$
NORMAL	0.877	0.687
DEFORMED	0.109	0.269
DEAD	0.014	0.044
Entropy, H	0.42	0.75

The boxplots alongside show that the uncertainty is appreciably larger for the 19 misclassified cases than for the 53 cases that the model classified correctly.

Model Uncertainty

All measurements, even the simplest, involve models, which can be deterministic or stochastic and are specified using mathematical or statistical constructs.

Deterministic models describe relations between quantities that are based on a physical theory. For example, the relation between pressure, density and velocity in Bernoulli's equation, which underlies the measurement of incompressible fluid flow.

Stochastic models, on the other hand, use probability distributions to describe relations between quantities that are influenced by natural variability or measurement uncertainty. For example, the relation between age, weight, and height in human population, which is determined via statistical data reductions.

Since building or selecting a measurement model is an integral part of measurement, and typically it is surrounded by uncertainty, this uncertainty contribution should be evaluated and propagated to the estimate of the measurand, the same as the contributions from all the other sources of measurement uncertainty.

Mass of Pluto

Modeling the motion of the heavenly bodies that comprise the solar system has fascinated scientists for centuries. As a feat of mathematical modeling and precision measurements, Neptune was discovered in 1846 based on the analysis of observational data about the motion of Uranus. This discovery remains one of the best examples of the power of the scientific method, and it prompted many at the time to look for the next planet that might lurk beyond the newly-discovered Neptune.

Four images from New Horizons Long Range Reconnaissance Imager were combined with color data from the Ralph instrument to create this global view of Pluto in July 2015 — Wikimedia Commons (NASA, 2015).

Already in 1848, well before Pluto's discovery, Jacques Babinet estimated the mass of a foretold new planet as 12 times that of Earth. Percival Lowell's 1915 prediction for "planet X" was 6.6 times Earth's mass. And when Clyde Tombaugh finally discovered it in 1930, the newspapers announced "a ninth planet, greater than earth, found." Only a few decades ago Pluto was thought to be several orders of magnitude heavier than we now know it to be. What happened that so drastically changed our estimates of Pluto's mass?

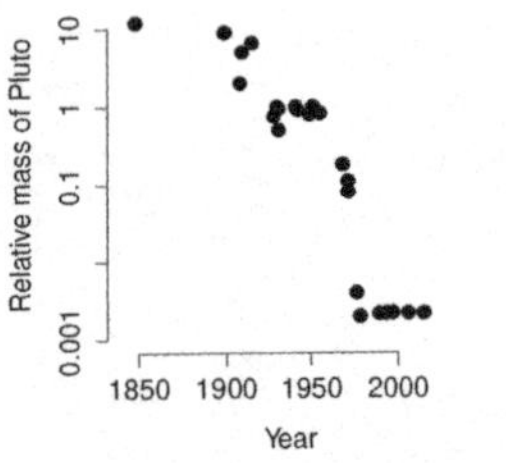

Estimated mass of Pluto ($m_{\text{Pluto}} / m_{\text{Earth}}$) over the last two centuries [Duncombe and Seidelmann, 1980] serves as a prime example of how important measurement models and all assumptions that go into these models are in creating knowledge.

Pluto is so distant that it is difficult to learn much about it from direct observation. Our knowledge of its mass therefore depends on the physical models we adopt. For a long time, Pluto's mass was estimated based on perturbations to the motions of Uranus and Neptune.

It all changed in 1978, when a sharp-eyed US astronomer, James W. Christy, discovered Pluto's first moon. At half the size of Pluto, Charon has a significant effect on Pluto's motion and enabled estimating its mass by application of Kepler's laws.

In the late 1980s, the orbits of Pluto and its largest moon Charon were aligned with the line-of-sight from Earth (an arrangement that occurs once in 120 years) which allowed for accurate mass estimates for the first time.[89] In 2015, NASA's *New Horizons* space probe flew near Pluto and was able to answer one of the most basic questions about Pluto conclusively, estimating its mass to be $m_{\text{Pluto}} = 0.0022 m_{\text{Earth}}$.

[89] R. P. Binzel. Pluto-Charon mutual events. *Geophysical Research Letters*, 16(11): 1205–1208, November 1989. doi:10.1029/gl016i011p01205

Similarly to the mass of Pluto, the mass of Sun had also been difficult to estimate in the past. The successive editions of Newton's *Principia* document such difficulties for the solar mass relative to Earth's mass whose contemporary value is approximately 330 000:

28 700	*Principia* (1687) 1st ed
227 512	*Principia* (1713) 2nd ed
169 282	*Principia* (1726) 3rd ed

Also note the excessive number of digits which serves as a reminder that even great men like Newton were not fully appreciative of measurement uncertainty.

Scientists tend to overestimate the confidence in their results and the quest for the mass of Pluto is not the only example where our collective scientific judgment has fallen short. Determinations of the atomic weights of tellurium and iodine made in the 19th century did not favor Mendeleev's suggestion [Mendeleev, 1871, Page 151][90] that the atomic weight of tellurium should be smaller than that of iodine. It is therefore not surprising that the estimates of these two atomic weights should have changed gradually to conform with Mendeleev's suggestion.[91]

[90] D. Mendeleev. Die periodische Gesetzmässigkeit der Elemente. *Annalen der Chemie und Pharmacie*, VIII. Suplementband:133–229, 1871

[91] J. Meija. Atomic weights of the elements: From measurements to the periodic table. In M. J.T. Milton, D. S. Wiersma, C. J. Williams, and M. Sega, editors, *New Frontiers for Metrology: From Biology and Chemistry to Quantum and Data Science*, volume 206 of *Proceedings of the International School of Physics "Enrico Fermi"*, pages 77–93. IOS Press, Amsterdam, The Netherlands, 2021. ISBN 978-1-64368-246-4. doi:10.3254/ENFI210019

Scientists tend to overestimate the confidence in their results and the quest for the mass of Pluto is not the only example where our collective scientific judgment has fallen short. Determinations of the atomic weights of tellurium and iodine made in the 19th century did not favor Mendeleev's suggestion that the atomic weight of tellurium should be smaller than that of iodine. It is therefore not surprising that the estimates of these two atomic weights should have changed gradually to conform with Mendeleev's suggestion.

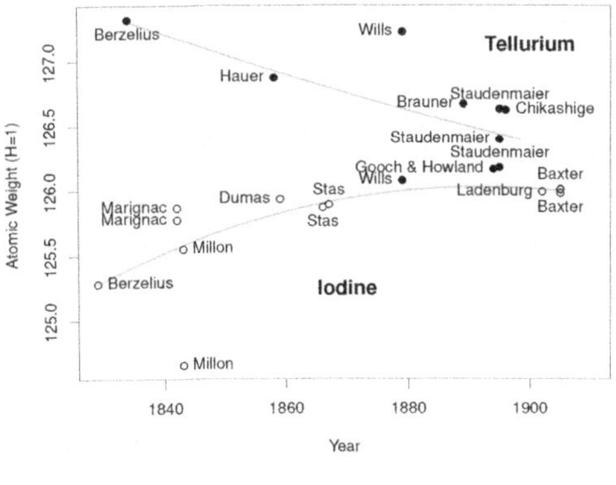

Each point represents the median of several determinations of the atomic weight of tellurium or iodine, made by an individual author based on a particular ratio of molecular weights.

For example, von Hauer [1857] made five determinations of the ratio between the molecular weights of AgBr and K_2TeBr_6, from which he could derive five estimates of the atomic weight of tellurium given values of the atomic weights of silver, bromine, and potassium that had been determined previously.

Note the temporal trends of the measured values of $A_r(\mathrm{Te})$ and $A_r(\mathrm{I})$, apparently reflecting a desire to gradually conform with Mendeleev's hypothesis.

Although now we are certain that the atomic weight of tellurium, 127.60 ± 0.03, is greater than that of iodine, $126.904\,47 \pm 0.000\,03$, it is plausible that Mendeleev's pronouncement played an invisible guiding role in contemporary atomic weight measurements of these two elements.

This phenomenon is known as the *expectation bias* and it is a reminder that uncertainty estimates are often influenced by unknown effects that have little to do with the measurement they pertain to.

Height of Mount Everest

Mount Everest: view from the south — Wikimedia Commons (shrimp01967, 2012).

Only in 1849, in the course of the Great Trigonometrical Survey of India (1802–1871), was Mount Everest recognized as the highest mountain on Earth.[92]

[92] S. G. Burrard. Mount Everest: The story of a long controversy. *Nature*, 71:42–46, November 1904. doi:10.1038/071042a0

The quest to measure the height of Mount Everest reveals how aspects of measurement models that are much too often hidden from view can influence the results. The earliest observations were made from northern India, some 160 km away, and involved measurements of angles made using theodolites.

Determinations of the height of Mount Everest extracted from the Records of the Great Trigonometrical Survey of India, based on observations made between November 1849 and January 1850 [Burrard, 1904].

STATION	DISTANCE	ANGLE	HEIGHT
Jirol	190.966 km	1° 53′ 33.35″	8836 m
Mirzapur	175.219 km	2° 11′ 16.66″	8841 m
Janjipati	174.392 km	2° 12′ 9.31″	8840 m
Ladnia	175.195 km	2° 11′ 25.52″	8839 m
Harpur	179.479 km	2° 6′ 24.98″	8847 m
Minai	183.081 km	2° 2′ 16.61″	8836 m

The simplest approach to estimate the height involves only the elevation angle (a), the distance from the observing station to the mountain (d), the altitude of the station (h_S), and a trigonometric relation:

$$h = h_S + d \tan a.$$

For the Jirol station, which stands 67 m above sea level, this formula yields

$$h = 67\,\text{m} + (190\,966\,\text{m}) \times \tan(1^\circ\,53'\,33.35'') \approx 6377\,\text{m},$$

which grossly underestimates the height of the mountain. If left uncorrected, Earth's curvature and the refraction of light as it travels through the atmosphere are the principal sources of error in trigonometric determinations of height made from long distances.

Accounting for the curvature (with Earth modeled as a sphere of radius $R = 6371$ km) leads to a much more complex model:

$$\sin(\tfrac{\pi}{2} - a) = \frac{(R+h)\sin(d/R)}{\sqrt{R_S^2 - 2R_S(R+h)\cos(d/R) + (R+h)^2}},$$

where $R_S = R + h_S$. Solving this equation for h numerically, again using the elevation angle measured from the Jirol station, gives $h \approx 9251$ m, now overestimating the height of Mount Everest.

Troughton & Simms theodolite from around 1910, used to measure angles in horizontal and vertical planes — Wikimedia Commons (Colgill, 2020).

The fact that atmospheric refraction tends to increase the apparent elevation angle of a mountain peak relative to the observer, is the main reason why the previous height estimate is biased high.

While atmospheric refraction depends on several environmental conditions, its magnitude is approximately 10 % of the effect of the Earth's curvature. The *Manual of Surveying for India* [Thuillier and Smyth, 1875, Page 505] explains how refraction was modeled:

> There are no fixed rules for Terrestrial refraction, but [...] in determining the heights of the peaks of the Snowy Range (Himalayas), about one-thirteenth of the contained arc was assumed.

The contained arc is the value (in radian) of the angle with vertex at the center of Earth subtended by an arc of length d on Earth's surface. It is the ratio of d to Earth's radius.

Thus, the effect of light refraction was modeled by reducing the observed elevation angle by $(d/R)/13$, that is from a to $a - (d/R)/13$ (expressed in radian). As a result, the estimate of the height of Mount Everest, still based on the observation made from Jirol, but now taking into account both Earth's curvature and atmospheric refraction, becomes $h \approx 8810$ m.

Other influences on the height estimates were recognized later, such as the effect of temperature on the refraction of light and the gravitational influence of these large mountains on plumb lines and leveling devices.

Despite all these challenges, the original average estimate from the 1850s, 8840 m, is remarkably close to the current estimate of 8848 m, based on GPS measurements made at the mountaintop.

In 1914, *Nature* noted that "when all is said and done, it is the errors arising from the deflection of the plumb-line [. . .], and the possible variation in the actual height of the point observed (common enough in the case of snow-capped peaks), which chiefly affect the accuracy of angular determinations of altitude, and it is probably to these [. . .] that we must ascribe [. . .] the doubt whether Kinchinjunga or K2 is to hold the honourable position of second in altitude to Everest amongst the world's highest peaks."

Significant efforts are devoted to this day in determining the precise height of Mount Everest. Over the last few decades, surveyors from China and Nepal, as well as researchers from other countries, have conducted independent measurement campaigns to measure the mountain. Not only were the estimates of the mountain's height different, not everyone even agreed on what type of height to use. The mountaintop has a 3.5 meter deep snow cap so it makes a big difference whether to use the rock height of the snow height.

After more than a decade of dispute, in December 2020 China and Nepal agreed on how tall Mount Everest. When the joint announcement of the new height was made by the representatives of both countries, Mount Everest grew taller by nearly a meter compared to the previous value! The new "official" value is 8848.86 m, and refers to the snow height of the mountain.

Averaging Measurement Models

In many measurement situations, several alternative models naturally present themselves, with no *a priori* reason to favor one over the others. In some cases it may be most convenient to select and use the "best" model among a collection of alternatives, like we did when we introduced a reliable guide for model building (Page 100) in the context of building a calibration function. In other cases, the best performance is achieved by a weighted average of alternative models.

In general, model averaging does not mean averaging the parameters of the alternative models. The alternative models may have different numbers of parameters, or, even if they have the same number of parameters, the parameters of different models may not be the same kinds of quantities that one could reasonably average. Instead, the averaging will be of predictions that the alternative models make of the same quantities, and the question is how to evaluate the uncertainty of such averages.

Influenza Epidemic

The following example illustrates model averaging to produce an estimate of the basic reproduction number (R_0) for an influenza epidemic that ravaged a boarding school for boys between the ages of 10 and 18 in the north of England, during January and February of 1978.

The concept of R_0 is often regarded to be one of the most useful tools in mathematical biology. It is the average number of infections produced by an infective person that interacts freely with others who are susceptible to becoming infected.

Measurement models for epidemics in human or animal populations typically comprise a deterministic component that describes the temporal evolution of the expected number of cases (and the corresponding expected numbers of individuals who are susceptible but not yet sick, of individuals who have already recovered, etc.) [Hethcote, 2000].

These models also comprise a stochastic component that describes how the actual counts of individuals in the different categories vary around their expected values [Bjørnstad, 2018].

DATE	CASES
1978-01-22	3
1978-01-23	8
1978-01-24	26
1978-01-25	76
1978-01-26	225
1978-01-27	298
1978-01-28	258
1978-01-29	233
1978-01-30	189
1978-01-31	128
1978-02-01	68
1978-02-02	29
1978-02-03	14
1978-02-04	4

English boarding school epidemic of 1978 [BMJ News and Notes, 1978; Martcheva, 2010].

The particular epidemic we will be concerned with started in late January and ended in early February of 1978, eventually infecting 512 of the 763 boys in the school. At the peak of the epidemic, 298 boys were confined to bed in the school's infirmary.

We will consider two mathematical models for the daily counts of influenza cases in the boarding school. Each involves a *compartment model* and Poisson (Page 173) random variables.

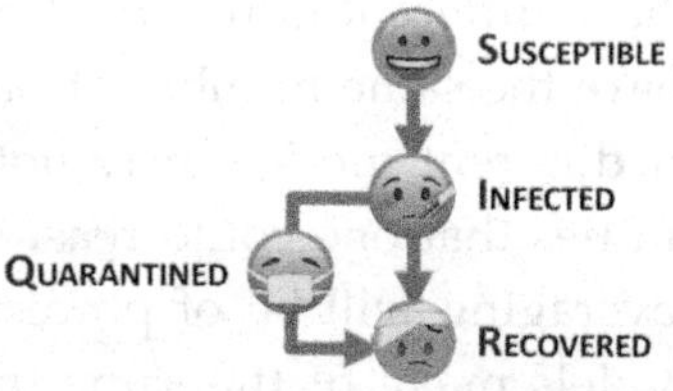

Schematics of two epidemiological, compartmental models of influenza. The SIR model considers only the *susceptible, infected,* and *recovered,* whereas the SIQR model considers also the *quarantined.*

At each epoch (a day in this case) a compartment model partitions the relevant population into several categories. For the SIR model these categories are the susceptible, the infective, and the recovered — whose initials, SIR, make the acronym of the model. The SIQR model comprises yet another category, the quarantined. The same person will belong to different categories at different times as the epidemic spreads and the disease progresses.

The SIR model was introduced in the 1920s [Kermack and McKendrick, 1927] and remains one of the simplest models for infectious diseases that are transmitted from human to human, and where recovery confers lasting resistance. This three-compartment model has undergone many improvements and additions tailored for a variety of situations. Recently, for example, the COVID-19 epidemic and the implementation of nationwide interventions in Italy were modeled using an extension of this model that comprises eight compartments: susceptible, infected, diagnosed, ailing, recognized, threatened, healed, and extinct [Giordano et al., 2020].

We will assume that, at the outset of the epidemic, exactly one boy is infective, and all the others are susceptible. Therefore, the initial counts (on day 1) in the different compartments are

$$S(1) = 762, \quad I(1) = 1, \quad Q(1) = 0, \quad R(1) = 0.$$

According to the SIR model, an infected boy will remain infective for some time, and then will recover, in the process acquiring immunity against reinfection with the same virus. But while he is infective, he continues to interact with the other boys in the school, likely spreading the disease.

This is not what actually happened: sick boys were isolated (that is, quarantined) in the school infirmary as soon as the obvious symptoms developed. Quarantining removed them from the pool of those that were spreading the disease. Regardless of whether a sick boy was quarantined or not, eventually he will recover. The SIQR model takes into account the effect of quarantining.

The deterministic components of the SIR and SIQR models are solutions of systems of differential equations, thus assuming that the numbers of boys in the different categories vary continuously over time. The three simultaneous differential equations for the deterministic component of the SIR model are

$$\begin{aligned} dS/dt &= -\beta SI/N, \\ dI/dt &= +\beta SI/N - \gamma I, \\ dR/dt &= +\gamma I, \end{aligned}$$

Since the time derivatives of the numbers of boys in the different compartments add to zero, the total $N = S + I + R$ remains constant over time. More complex models can take into account births and deaths (regardless of whether these are caused by the disease).

where $N = 763$ is the total number of boys in the school. Note that S, I and R all are functions of time, t, even if this is not shown explicitly.

The observations are the numbers of boys that are sick in bed on each day of the epidemic, which are modeled as outcomes of independent Poisson random variables with means $I(1), \ldots, I(14)$. If the variability of these counts were much in excess of $\sqrt{I(1)}, \ldots, \sqrt{I(14)}$, then a negative binomial (Page 175) model might be preferable.

The SIQR model has an additional parameter, α, which is the quarantining rate. We assume that the same recovery rate γ applies to all infectives, regardless of whether they are quarantined or not. The SIQR model is represented by the following system of four differential equations:

$$\begin{aligned} dS/dt &= -\beta SI/(N-Q), \\ dI/dt &= +\beta SI/(N-Q) - \gamma I - \alpha I, \\ dQ/dt &= +\alpha I - \gamma Q, \\ dR/dt &= +\gamma I + \gamma Q. \end{aligned}$$

These two epidemiological models were fitted to the data using the Stan modeling language, in tandem with the R package `rstan`.[93] The estimates of all non-observable quantities are the means of their Bayesian posterior distributions (Page 204).

[93] B. Carpenter, A. Gelman, M. Hoffman, D. Lee, B. Goodrich, M. Betancourt, M. Brubaker, J. Guo, P. Li, and A. Riddell. Stan: A probabilistic programming language. *Journal of Statistical Software*, 76(1):1–32, 2017. doi:10.18637/jss.v076.i01; and Stan Development Team. *Stan User's Guide*. `mc-stan.org`, 2019. Stan Version 2.28

The following Stan code was used to fit the SIR model, assuming that the counts of boys in the different compartments are like outcomes of Poisson random variables whose means satisfy the corresponding system of differential equations presented above.

```
modelSIR = "
functions {
  real[] sir(real t, real[] y, real[] ps, real[] xr, int[] xi)
  { real N = xi[1];
    real dSdt = - ps[1] * y[1] * y[2] / N;
    real dIdt =   ps[1] * y[1] * y[2] / N - ps[2] * y[2];
    real dRdt =   ps[2] * y[2];
    return {dSdt, dIdt, dRdt}; }
 }
data { int N; real y0[3]; real ts[14]; int cases[14]; }
transformed data { real xr[0]; int xi[1] = N; }
parameters { real<lower=0> bg[2]; }
transformed parameters {
      real y[14,3];
      y = integrate_ode_rk45(sir, y0, 0, ts, bg, xr, xi);
 }
model {
      bg ~ normal(1, 10);          // Priors for beta and gamma
      cases ~ poisson(y[,2]);   // Sampling distribution
 }
generated quantities {
      real R0 = bg[1] / bg[2];
 }"
```

The R code below compiles and fits the Stan model to the data. Similar codes were used to fit the SIQR model.

```
library(rstan)
library(outbreaks)
cases = influenza_england_1978_school$in_bed
N = 763; n_days = length(cases)
dataSIR = list(n_days=n_days, y0 = c(S=N-1, I=1, R=0),
               N = N, cases = cases, t0 = 0, ts = seq(1, n_days),
               ts_pred = seq(1, 1+n_days, length.out = 100) )
## Compile the Stan model
modelSIR.poisson = stan_model(model_code=modelSIR)
## Fit the Stan model
fitSIR.poisson = sampling(modelSIR.poisson, data = dataSIR)
## Estimate of R0
print(fitSIR.poisson, pars = 'R0')
```

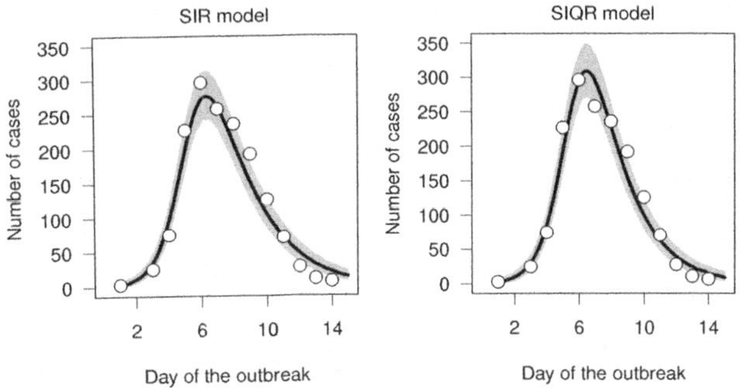

Observed daily numbers of cases and corresponding predicted counts produced by SIR and SIQR models with Poisson variability on the observed cases, surrounded by 95% probability bands.

The basic reproduction numbers for the SIR and SIQR models are given as follows:

$$R_0(\text{SIR}) = \frac{\beta}{\gamma},$$

$$R_0(\text{SIQR}) = \frac{\beta}{\gamma + \alpha}.$$

The Bayesian estimates of R_0 along with their standard uncertainties are given in the alongside table. Although numerically different, they are not significantly different once their associated uncertainties are taken into account: their standardized difference is

MODEL	R_0	$u(R_0)$
SIR	3.55	0.08
SIQR	3.38	0.08

$$z = \frac{3.55 - 3.38}{\sqrt{0.08^2 + 0.08^2}} = 1.5.$$

Since these Bayesian estimates are approximately like

outcomes of Gaussian random variables, a z-test for their difference yields p-value 0.13 (Page 32).

R_0 captures various aspects of the outbreak. For simple models such as these, the proportion of the population that needs to be immunized to prevent sustained spread of the disease (that is, to achieve *herd immunity*), has to be larger than $1 - 1/R_0$ and the maximum number of cases on any given day is $I_{\max} = N - N(1 + \ln R_0)/R_0$.

The estimates of R_0 produced by these two models can be averaged using Bayesian *stacking weights* [Yao et al., 2017] to produce an estimate corresponding to the best mixture of these models. The weights were computed using R package `loo` [Vehtari et al., 2019]. Since the *stacking weights* were 0.24 for SIR and 0.76 for SIQR, the combined estimate is

$$R_0 = (0.24 \times 3.55) + (0.76 \times 3.38) = 3.42,$$

with standard uncertainty

$$u(R_0) = \sqrt{(0.24 \times 0.08)^2 + (0.76 \times 0.08)^2} = 0.06.$$

We have here an instance of the magic of averaging: the uncertainty surrounding the weighted average of the estimates of R_0 corresponding to the SIR and SIQR models, is smaller than the uncertainties of these estimates.

The basic reproduction number, R_0, represents the average number of new infections per existing case. In other words, if $R_0 = 3$, then one person with the disease is expected to infect, on average, three others. Despite its simplicity, R_0 is a *messy* quantity because the definition allows for a multitude of interpretations. For example, do we estimate this quantity at the beginning of the outbreak, at the end, or somehow estimate the average during the entire infectious period?

For measles, R_0 is widely believed to be somewhere between 12 and 18. Yet, as an example of the real-world *messiness* of the R_0 estimates, a recent systematic review of 18 studies of measles outbreaks reported R_0 values ranging from 4 to 200 [Guerra et al., 2017].

A common way to estimate R_0, among the many available alternatives,[94] is based on the total number of susceptible patients at the end of the outbreak, which for the boys school was $S(\infty) = 763 - 512 = 251$:

$$R_0 = \frac{\ln(S(0)/S(\infty))}{1 - S(\infty)/N} = \frac{\ln(762/251)}{1 - 251/763} = 1.65.$$

[94] J. M. Heffernan, R. J. Smith, and L. M. Wahl. Perspectives on the basic reproductive ratio. *Journal of The Royal Society Interface*, 2:281–293, 2005. doi:10.1098/rsif.2005.0042

Although *ad hoc*, rather than model-based as the estimates computed above, the very fact that this estimate of R_0 differs from them to such enormous extent highlights the role of models, and reveals the impact that the selection of a model has upon the uncertainty associated with estimates of the quantities of interest.

Age of Machu Picchu

A recent study[95] concluded that Machu Picchu is older than previously thought. This study applied radiocarbon dating to human bone and tooth samples retrieved from burial caves at Machu Picchu. For one of the oldest molar specimens, radiocarbon dating yielded an age of 540 BP with standard uncertainty 20 years.

Machu Picchu older than expected, study reveals Mike Cummings (August 4th, 2021) *YaleNews*

[95] R. L. Burger, L. C. Salazar, J. Nesbitt, E. Washburn, and L. Fehren-Schmitz. New AMS dates for Machu Picchu: results and implications. *Antiquity*, 95(383):1265–1279, 2021. doi:10.15184/aqy.2021.99

To convert carbon-14 levels into the corresponding calendar ages, one uses either the Northern Hemisphere Curve (INTCAL2020) or the Southern Hemisphere Curve (SHCAL2020). While Machu Picchu is located in the Southern hemisphere, due to its proximity to the equator and local weather patterns, it gets its carbon from both hemispheres.[96] Since the precise nature of atmospheric mixing is unknown, Burger *et al.* (2021) linearly pooled these two curves to recognize the carbon sources from both hemispheres.

[96] E. J. Marsh, M. C. Bruno, S. C. Fritz, P. Baker, J. M. Capriles, and C. A. Hastorf. IntCal, SHCal, or a Mixed Curve? Choosing a ^{14}C calibration curve for archaeological and paleoenvironmental records from tropical South America. *Radiocarbon*, 60(3):925–940, 2018. doi:10.1017/RDC.2018.16

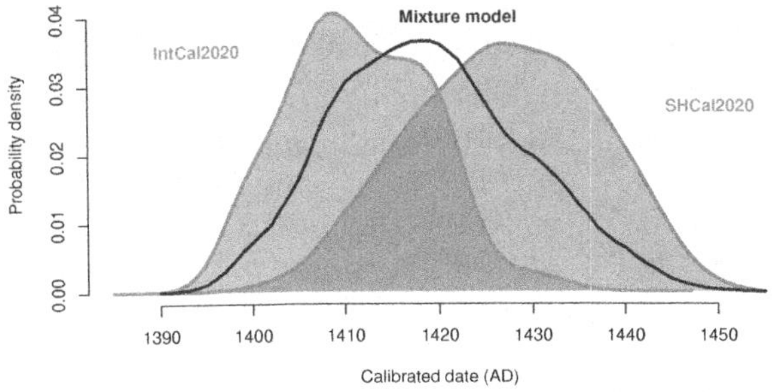

Comparison of the calibrated radiocarbon dates for molar 4F listed in Figure 3 and Table S5 of Burger *et al.* (2021) obtained using the calibration curves SHCAL2020, INTCAL2020, and their mixture.

The results showed that Machu Picchu was occupied from as early as AD 1420, two decades earlier than suggested by the textual sources that associate the site with Emperor Pachacuti's rise to power in AD 1438. The ages of molar 4F corresponding to the SHCAL2020 and INTCAL2020 calibration curves differ by 16 years: thus, model uncertainty emerges as a source of uncertainty at least as important as the model-specific standard measurement uncertainty, which is around 12 years.

Consensus Building

Burgess and Spangler [2003] explain that "consensus building (also known as collaborative problem solving or collaboration) is a conflict-resolution process used mainly to settle complex, multiparty disputes." In the sciences, consensus building serves to blend measurement results for the same measurand that have been obtained independently of one another.

In measurement science in particular, besides this role, consensus building is also used to characterize and compare the different measurement results, by estimating the difference between the true value that each purports to measure, and the true value of the consensus value, and evaluating the corresponding uncertainty — the so-called *degrees of equivalence* [Koepke et al., 2017].

In medicine, where consensus building is often referred to as *meta-analysis* [Higgins et al., 2019], and where the same techniques are also employed to merge results of multicenter trials [Friedman et al., 2015], the goal is to ascertain confidently that a medical procedure or therapy is superior to another, by pooling results from different studies that, if taken individually, may be inconclusive.

Hubble-Lemaître Constant

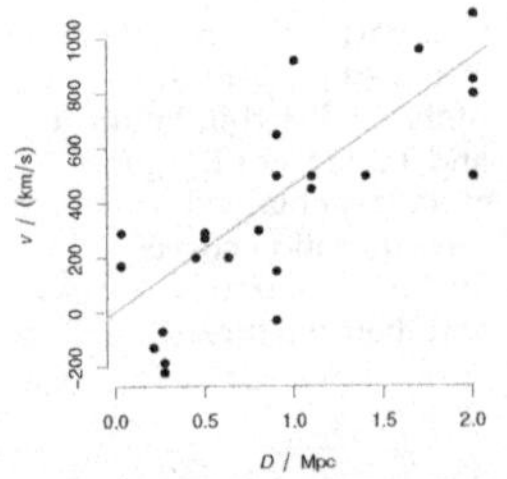

Measured values of distance and velocity for 24 galaxies reported by Hubble [1929, Table 1]. The gray line has a zero intercept and Hubble's estimate for the slope, $465 \pm 50\,(\text{km/s})/\text{Mpc}$.

In the 1920s, Edwin Hubble and Georges Lemaître discovered that galaxies appear to be moving away from Earth at speeds (v) that are proportional to their distance (D) from Earth [Hubble, 1929] [Lemaître, 1927, 2013]:

$$v = H_0 D.$$

The constant of proportionality, H_0, is known as the Hubble-Lemaître constant. This discovery motivated Einstein to visit Hubble at the Mount Wilson observatory on January 29, 1931, and acknowledge that the universe indeed is expanding.

Since the final release of the results from the *Planck* survey,[97] which include an estimate of H_0, several other measurement results have been produced for this constant, by application of a wide variety of methods. The measurement results are mutually inconsistent in the sense that the measured values are more dispersed than their uncertainties suggest that they should be.

[97] Planck Collab. et al. Planck 2018 results — VI. Cosmological parameters. *Astronomy & Astrophysics*, 641: A6, 2020. doi:10.1051/0004-6361/201833910

H_0	$u(H_0)$	STUDY	REFERENCE
67.36	0.54	PLANCK	[Planck Collab. et al., 2020]
72.50	2.20	HOLI	[Birrer et al., 2019]
67.80	1.30	DES	[Macaulay et al., 2019]
69.32	1.42	RYAN	[Ryan et al., 2019]
74.03	1.42	HST	[Riess et al., 2019]
67.40	6.10	FLAT	[Domínguez et al., 2019]
70.30	5.15	LV	[Hotokezaka et al., 2019]
73.30	1.75	HOL6	[Wong et al., 2019]
69.80	1.90	HST	[Freedman et al., 2019]
73.50	1.40	RPR	[Reid et al., 2019]
70.30	1.35	DUTTA	[Dutta et al., 2019]
76.80	2.60	SH3	[Chen et al., 2019]
74.20	2.85	STRI	[Shajib et al., 2019]
73.90	3.00	MEGA	[Pesce et al., 2020]
75.80	5.05	JAEG	[de Jaeger et al., 2020]
67.60	4.25	MUK	[Mukherjee et al., 2020]
71.80	3.59	DENZ	[Denzel et al., 2020]
73.50	5.30	BAX	[Baxter and Sherwin, 2020]
67.40	1.00	SEDG	[Sedgwick et al., 2021]
73.20	1.30	HSTGA	[Riess et al., 2021a]
72.10	2.00	HSTGB	[Soltis et al., 2021]
69.80	1.70	FREE	[Freedman, 2021]
73.30	1.40	SHOES	[Riess et al., 2021b]

Selected, recent estimates of the Hubble-Lemaître constant, H_0 in (km/s)/Mpc. The uncertainties listed in the column headed $u(H_0)$ are such that each of the intervals $\{H_{0,j} \pm u(H_{0,j})\}$ is believed (by its authors) to include the true value of H_0 with probability 68 % approximately. Some of the uncertainties were originally expressed asymmetrically, but since the asymmetries were very mild, here they have been replaced by the geometric averages of the corresponding, reported "left" and "right" uncertainties.

In particular, there is a statistically significant discrepancy between the estimates of H_0 based on recent measurements of the velocity and distance of galaxies, and the estimate derived from measurements of fluctuations and polarization of the cosmic microwave background made by the *Planck* mission, which ended in 2013.

[98] V. Poulin, T. L. Smith, T. Karwal, and M. Kamionkowski. Early dark energy can resolve the Hubble tension. *Physical Review Letters*, 122:221301, June 2019. doi:10.1103/PhysRevLett.122.221301

Astrophysicists call this discrepancy the *Hubble tension*,[98] noting that "the significance of the current tension also depends on the assumption that all sources of uncertainty have been recognized and accounted for."[99]

[99] W. L. Freedman. Measurements of the Hubble Constant: Tensions in perspective. *The Astrophysical Journal*, 919(1):16, 2021. doi:10.3847/1538-4357/ac0e95

While remaining agnostic about the origin of the aforementioned "excess" dispersion, it is possible to blend mutually inconsistent measurement results and produce a single *consensus* estimate, using this statistical measurement model:

$$H_{0,j} = H_0 + \lambda_j + \varepsilon_j \quad (j = 1, \dots, n),$$

where the $\{\lambda_j\}$ denote experiment effects, and the $\{\varepsilon_j\}$ denote experiment-specific measurement errors.

Both the experiment effects and the measurement errors are assumed to have mean zero, thus conveying the (debatable) assumption that the measured values, taken as a group, are centered at the correct, true value of H_0. The $\{\varepsilon_j\}$ have standard deviations equal to the reported uncertainties listed under $u(H_0)$ in the table above.

The $\{\lambda_j\}$ have standard deviation τ that quantifies the aforementioned "excess" dispersion, which manifests itself only when results from multiple, independent experiments are compared. For this reason, τ is often called *dark uncertainty* [Thompson and Ellison, 2011].

The standard deviation of the 22 estimates of the Hubble-Lemaître constant listed above (excluding the value from *Planck*) is 2.76 (km/s)/Mpc, while the median of the corresponding, reported standard uncertainties is 1.95 (km/s)/Mpc. Thus, a preliminary estimate of τ is

$$\tau = \sqrt{2.76^2 - 1.95^2} = 1.95 \text{ (km/s)/Mpc}.$$

Further to specify the model for the $\{H_{0,j}\}$, we will employ the *NIST Decision Tree*,[100] which recommends that the $\{\lambda_j\}$ be modeled as a sample from a Laplace distribution (Page 172), which can accommodate large deviations from the mean more naturally than the Gaussian distribution can. The statistical model assumes that the standard uncertainties $\{u(H_{0,j})\}$ are based on large numbers of degrees of freedom, hence treats them as known constants.

[100] A. Possolo, A. Koepke, D. Newton, and M. R. Winchester. Decision tree for key comparisons. *Journal of Research of the National Institute of Standards and Technology*, 126:126007, 2021. doi:10.6028/jres.126.007

The measurement results, and the consensus value and its associated uncertainty are depicted below.

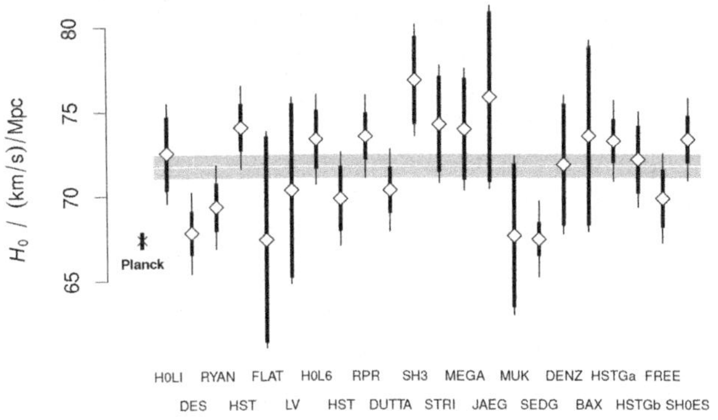

Each diamond represents a measured value, each vertical, thick line segment represents an interval $H_{0,j} \pm u(H_{0,j})$, and each vertical, thin line segment represents an interval $H_{0,j} \pm (\tau^2 + u^2(H_{0,j}))^{½}$. The horizontal, white line segment represents the consensus value derived from all measurement results except *Planck*'s, and the horizontal, shaded rectangle depicts the standard uncertainty associated with the consensus value.

[101] B. Carpenter, A. Gelman, M. Hoffman, D. Lee, B. Goodrich, M. Betancourt, M. Brubaker, J. Guo, P. Li, and A. Riddell. Stan: A probabilistic programming language. *Journal of Statistical Software*, 76(1):1–32, 2017. doi:10.18637/jss.v076.i01

A Bayesian version of the foregoing statistical measurement model was fitted to the measurement results using the Stan[101] and R codes listed below. The R code assumes that the Stan code has been assigned to variable `HLG.model` as a character string.

The median of the posterior distribution of τ is an estimate of the dark uncertainty: $\widehat{\tau} = 2.03\,(\text{km/s})/\text{Mpc}$, and the mean and standard deviation of the posterior distribution of the consensus value are

$$\widehat{H}_0 = 71.69\,(\text{km/s})/\text{Mpc}, \quad \text{and}$$
$$u(\widehat{H}_0) = 0.66\,(\text{km/s})/\text{Mpc}.$$

```
library(rstan)
H0.x = c(72.5, 67.8, 69.32, 74.03, 67.4, 70.3, 73.3, 69.8,
         73.5, 70.3, 76.8, 74.2, 73.9, 75.8, 67.6, 67.4,
         71.8, 73.5, 73.2, 72.1, 69.8, 73.3)
H0.ux = c(2.2, 1.3, 1.42, 1.42, 6.1, 5.15, 1.75, 1.9, 1.4,
          1.35, 2.6, 2.85, 3, 5.05, 4.25, 1, 3.59, 5.3,
          1.3, 2, 1.7, 1.4)

H0.Data = list(n=length(H0), x=H0.x, u=H0.ux,
               H0PriorMean=67.4, H0PriorSD=100, gamma=mad(H0.x))
H0.Fit = stan(model_code=HLG.model, data=H0.Data,
              warmup=500000, iter=1000000, chains=4, thin=25)
H0.Fit.post = extract(H0.Fit)

round(c(mean(H0.Fit.post$H0), sd(H0.Fit.post$H0)), 2)
round(median(H0.Fit.post$tau), 2)
```

```
HLG.model = "
data {
  int  n;                        // Number of measurement results
  real<lower=0> H0PriorMean; // Prior mean for mu
  real<lower=0> H0PriorSD;   // Prior SD for mu
  real gamma;                    // Prior median for tau
  vector<lower=0>[n] H0x;      // Measured values
  vector<lower=0>[n] H0ux;     // Standard uncertainties
}
parameters {
  real<lower=0> H0;              // True consensus value
  real<lower=0> tau;             // Dark uncertainty
  vector<lower=0>[n] theta;  // True values of H0x
}
model {
  // Prior for mu
  H0 ~ normal(H0PriorMean, H0PriorSD);
  // Half-Cauchy prior for tau with median gamma
  tau ~ cauchy(0, gamma);
  // Random effects {lambda[j] = theta[j] - H0}
  // Division by sqrt(2) makes tau the prior SD for {theta[j]}
  theta ~ double_exponential(H0, tau/sqrt(2));
  // Likelihood
  H0x ~ normal(theta, H0ux);
}"
```

The *Hubble time,* which is the reciprocal of the Hubble-Lemaître constant, corresponding to the foregoing consensus value $\widehat{H}_0$, suggests 13.7 billion years for the age of the universe:

$$t_H = \frac{1}{H_0} = \frac{\dfrac{3.1 \times 10^{19}\,\mathrm{km/Mpc}}{71.69\,(\mathrm{km/s})/\mathrm{Mpc}}}{31\,557\,600\,\mathrm{s/a}} \approx 13.70 \times 10^9\,\mathrm{a},$$

with standard uncertainty $u(t_H) = 0.13 \times 10^9$ a.

The value of 67.36(54) (km/s)/Mpc listed above is the estimate of H_0 that the Planck Collaboration (2018) declares to be the "best estimate" assuming the base ΛCDM cosmological model.[102]

[102] Planck Collab. et al. Planck 2018 results — VI. Cosmological parameters. *Astronomy & Astrophysics*, 641:A6, 2020. doi:10.1051/0004-6361/201833910

To compare the cosmological estimate derived from the *Planck* survey with the foregoing consensus value, we compute the normalized difference

$$z = \frac{71.69 - 67.36}{\sqrt{0.66^2 + 0.54^2}} = 5.1.$$

On the hypothesis of no difference between the corresponding true values, this normalized difference would be like an outcome of a Gaussian random variable with mean 0 and standard deviation 1. The probability of attaining or exceeding such a difference (regardless of sign) is $p = 4 \times 10^{-7}$, thus suggesting a very significant difference.

This discrepancy, which is an expression of the aforementioned *Hubble tension*, suggests that the pattern of expansion of the universe may have been somewhat more complex than the Hubble-Lemaître "law" contemplates, and indeed may lead to new physics.[103]

[103] J. Sokol. A recharged debate over the speed of the expansion of the universe could lead to new physics. *Science*, March 2017. doi:10.1126/science.aal0877

Arsenic in Kudzu

Kudzu comprises several species of perennial twining vines native to East Asia, which were introduced into the United States in 1876, originally intended as ornamental plants, and subsequently also used as food for cattle and ground cover.

Kudzu, "the vine that ate the South." — Kerry Britton, USDA Forest Service, Bugwood.org.

Their astonishingly rapid growth and ability to climb and develop roots opportunistically have turned kudzu into a damaging infestation, snuffing other plants large and small, including trees, and covering man-made structures.

The development of NIST SRM 3268 *Pueraria montana* var. *lobata* (Kudzu) Extract, included an interlaboratory study where 22 laboratories made triplicate determinations of the mass fraction of arsenic in this material, listed and depicted below, on the margin.

The Shapiro-Wilk test of Gaussian shape offers no compelling reason to abandon the hypothesis that all triplets are like samples from Gaussian distributions.

Therefore, the triplets will be replaced by their corresponding averages $\{w_j\}$ and associated standard uncertainties $\{u(w_j)\}$ evaluated using the Type A method from the GUM. For example, for laboratory V,

$$w_V = \frac{0.873 + 0.881 + 0.916}{3} = 0.890\,\text{mg/kg},$$

$$u^2(w_V) = \frac{(0.873 - w_V)^2}{3-1} + \frac{(0.881 - w_V)^2}{3-1} + \frac{(0.916 - w_V)^2}{3-1} = (0.013\,\text{mg/kg})^2.$$

A	0.922	0.978	0.988
B	0.531	0.545	0.535
C	0.908	0.928	0.911
D	1.376	1.399	1.388
E	0.899	0.852	0.830
F	1.008	1.004	1.002
G	0.912	0.912	0.922
H	0.892	0.886	0.857
I	0.793	0.794	0.801
J	1.027	1.030	1.044
K	0.913	0.957	0.956
L	0.747	0.795	0.823
M	0.978	1.015	0.936
N	0.779	0.956	1.026
O	0.952	0.949	0.948
P	0.851	0.866	0.871
Q	0.702	0.702	0.723
R	0.608	0.587	0.576
S	0.686	0.595	0.649
T	0.947	0.982	0.945
U	0.838	0.817	0.828
V	0.873	0.881	0.916

Triplicate determinations of the mass fraction (mg/kg) of arsenic in kudzu, made by laboratories A, B, ..., V.

Cochran's *Q*-test [Cochran, 1954] suggests that the sets of determinations are mutually inconsistent, or heterogeneous, in the sense that the averages of the triplets are significantly more dispersed than the individual determinations within the triplets. Such mutual inconsistency would still be present even if the determinations made by laboratories B, D, Q, R, and S were to be left out. However, they will not be left out from our subsequent analyses because there is no substantive reason to.

The symmetry test proposed by Miao et al. [2006] and implemented in R package `symmetry` [Ivanović et al., 2020], applied to the averages of the triplicates obtained by the participating laboratories, yields p-value 0.37 (Page 32), hence no reason to dismiss a symmetrical model for the random effects.

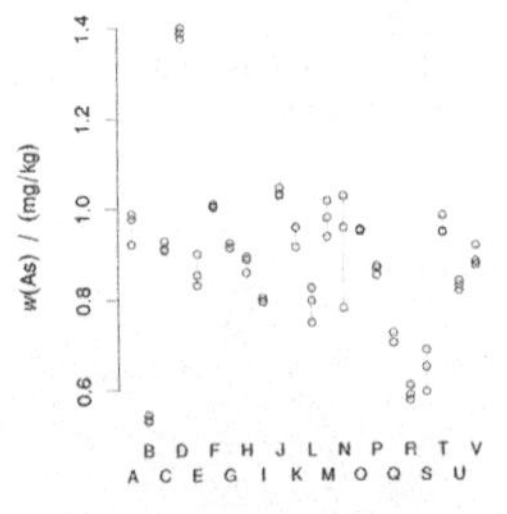

Each open circle represents a measured value, and each vertical line segment links the replicates from one laboratory.

And the Anderson-Darling test of Gaussian shape, applied to the coarsely standardized laboratory-specific averages, yields p-value 0.004 (Page 32). The "coarsely standardized" averages are $\{(w_j - M)/u(w_j)\}$, where M denotes the median of the $\{w_j\}$, and each w_j is the average of the three replicates obtained by laboratory j, for $j = \text{A}, \dots, \text{V}$.

Thus, we are faced with a situation where the laboratory-specific lack of repeatability may be modeled using Gaussian distributions, but the laboratory effects require a model that is symmetrical and has tails heavier than Gaussian tails, in particular to accommodate the results from laboratories B and D.

Considering the results of the foregoing statistical tests,

which underlie the recommendations for model selection that the *NIST Decision Tree* makes,[104], we will adopt a random effects model of the form $w_j = \omega + \lambda_j + \varepsilon_j$, where ω denotes the true value of the mass fraction of arsenic in the material, the $\{\lambda_j\}$ have a Laplace distribution (Page 172) with mean 0 and standard deviation τ, and the $\{\varepsilon_j\}$ are Gaussian, all with mean 0 but possibly different standard deviations $\{\sigma_j\}$.

[104] A. Possolo, A. Koepke, D. Newton, and M. R. Winchester. Decision tree for key comparisons. *Journal of Research of the National Institute of Standards and Technology*, 126:126007, 2021. doi:10.6028/jres.126.007

We will fit a Bayesian version of this model using the following code in the Stan language:[105]

[105] B. Carpenter, A. Gelman, M. Hoffman, D. Lee, B. Goodrich, M. Betancourt, M. Brubaker, J. Guo, P. Li, and A. Riddell. Stan: A probabilistic programming language. *Journal of Statistical Software*, 76(1):1–32, 2017. doi:10.18637/jss.v076.i01

```
data { int  n;      // Number of labs
  real gamma;     // Prior median of tau
  real delta;     // Prior median of {sigma2[j]}
  vector<lower=0>[n] w;     // Measured values
  vector<lower=0>[n] uw;    // Standard uncertainties
  vector<lower=2>[n] nu; } // Numbers of degrees of freedom

transformed data{ vector<lower=0>[n] uw2 = square(uw); }

parameters { real omega; // True overall mean
  real<lower=0> tau;      // Dark uncertainty
  vector<lower=0>[n] theta; // True lab means
  vector<lower=0>[n] sigma; } // True lab SDs

model { // Non-informative prior for omega
  omega ~ normal(0, 100);
  // Half-Cauchy prior for tau with median gamma
  tau ~ cauchy(0, gamma);
  // Random effects {lambda[j] = theta[j]-omega}
  // Division by sqrt(2) makes tau the prior SD
  theta ~ double_exponential(omega, tau/sqrt(2));
  // Half-Cauchy prior for sigmas with median delta
  sigma ~ cauchy(0, delta);
  // Likelihood for uw2
  for (j in 1:n) {uw2[j] ~ gamma(nu[j]/2,
                               nu[j]/(2*(sigma[j]^2)));}
  // Likelihood for w
  w ~ normal(theta, sigma); }
```

The Stan code treats both the measured values $\{w_j\}$ and the associated uncertainties $\{u_j\}$ as data. Therefore, the likelihood (Page 192) includes a term for the $\{u_j^2\}$ that recognizes the fact that, under the Gaussian assumption for the measured values, the $\{\nu_j u_j^2/\sigma_j^2\}$ are like outcomes of independent random variables with chi-square distributions (Page 170) with $\{\nu_j\}$ degrees of freedom, which happen to be all equal to 2.

In the model introduced above, $w_j = \omega + \lambda_j + \varepsilon_j$, the $\{u(w_j)\}$ are estimates of the $\{\sigma_j\}$ based on only 2 degrees of freedom each, and an estimate of τ will be based on the dispersion of only 22 observations. The Bayesian formulation is quite capable of recognizing such limitations and take them into account while evaluating the uncertainty associated with the consensus value.

Our Bayesian model includes the following prior distributions: a largely non-informative, Gaussian prior distribution for ω, a half-Cauchy (Page 170) prior distribution for τ, with median γ, and a half-Cauchy prior distribution for the $\{\sigma_j\}$, with median δ.

Since γ and δ are parameters of prior distributions, they are often called *hyperparameters*. We set γ equal to the mad of the laboratory-specific averages, and δ equal to the median of the $\{u(w_j)\}$.

The following R code executes the Stan code listed above after it will have been assigned to variable HLGS.model as a character string, including the line breaks, where w and uw are vectors of laboratory averages and associated standard uncertainties, and nu is the corresponding vector of numbers of degrees of freedom (whose 22 elements all should be equal to 2).

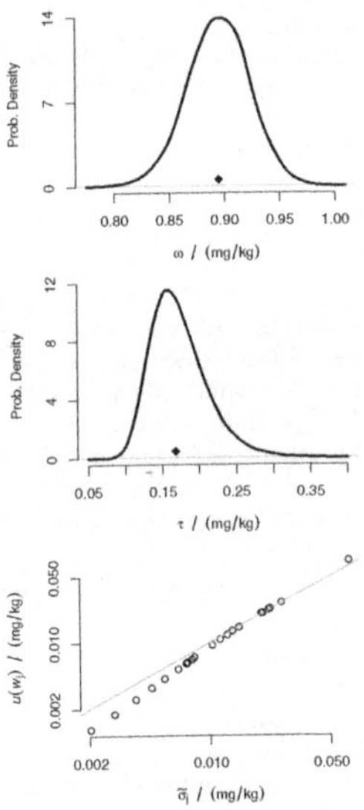

TOP PANEL: Posterior probability density of the consensus value, with mean $\tilde{\omega} = 0.895$ mg/kg (diamond). MIDDLE PANEL: Posterior probability density of the dark uncertainty, with median $\tilde{\tau} = 0.167$ mg/kg (diamond). BOTTOM PANEL: Reported standard uncertainties, $\{u(w_j)\}$, versus posterior medians of the corresponding $\{\sigma_j\}$.

```
library(rstan)
w = c(0.963, 0.537, 0.916, 1.388, 0.86, 1.005, 0.915, 0.878,
      0.796, 1.034, 0.942, 0.788, 0.976, 0.92, 0.95, 0.863,
      0.709, 0.59, 0.643, 0.958, 0.828, 0.89)
uw = c(0.021, 0.004, 0.006, 0.007, 0.02, 0.002, 0.003, 0.011,
       0.003, 0.005, 0.015, 0.022, 0.023, 0.073, 0.001, 0.006,
       0.007, 0.009, 0.026, 0.012, 0.006, 0.013)
nu = rep(2, length(w))
As.Data = list(n=length(w), w=w, uw=uw, nu=nu,
               gamma=mad(w), delta=median(uw))
As.Fit = stan(model_code=HLGS.model, data=As.Data,
              warmup=500000, iter=1000000,
              control=list(adapt_delta=0.975),
              chains=4, cores=4, thin=25)
print(As.Fit, digits=3)
```

An estimate of the posterior probability density (Page 159) of the consensus value ω is depicted in the top panel, alongside. The posterior mean is $\tilde{\omega} = 0.895$ mg/kg, whose associated uncertainty is $u(\tilde{\omega}) = 0.029$ mg/kg. A 95 % credible interval for ω ranges from 0.839 to 0.951 mg/kg.

The bottom panel of the figure alongside shows that posterior median uncertainties tend to be larger than the reported uncertainties for the smaller uncertainties, and smaller than them for the larger uncertainties: a shrinkage effect that is typical of Bayesian estimates.

The posterior median for the dark uncertainty, $\tilde{\tau} = 0.167\,\mathrm{mg/kg}$ is about 20 times larger than the median of the uncertainties $\{u(w_j)\}$, $0.008\,19\,\mathrm{mg/kg}$. Accordingly, dark uncertainty makes a much larger contribution to the uncertainty associated with the consensus value than the Type A uncertainty evaluations for the laboratory averages.

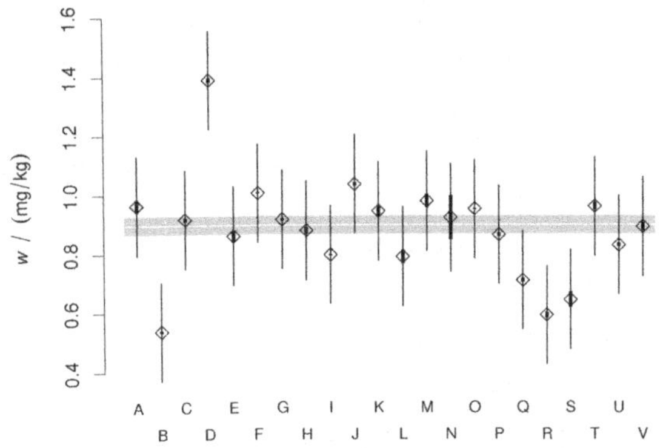

Each diamond represents a measured value, each vertical, thick line segment represents an interval $w_j \pm u(w_j)$, and each vertical, thin line segment represents an interval $w_j \pm (\tau^2 + u^2(w_j))^{1/2}$. The horizontal, white line segment represents the consensus value, and the horizontal, shaded rectangle depicts the standard uncertainty associated with the consensus value.

The following one-liner uses facilities implemented in R package `brms`[106] package for R, and produces nearly identical results. Since the model implemented in `brm` and the model specified above differ only in the choice of prior distributions, the fair agreement of the respective results is a welcome outcome of this sensitivity analysis. `As` denotes an R data frame with $22 \times 3 = 66$ rows and two columns: one (named `lab`) has the laboratory labels, and the other (named `w`) has the replicated determinations of the mass fraction of arsenic made by the 22 laboratories:

[106] P.-C. Bürkner. Advanced Bayesian multilevel modeling with the R package brms. *The R Journal*, 10(1):395–411, 2018. doi:10.32614/RJ-2018-017

```
library(brms)
brm(formula = bf(w ~ 1 + 1|lab, sigma ~ 0 + lab, quantile=0.5),
    family = asym_laplace, data = As)
```

Appendix: Uncertainty

Uncertainty is judged by some to be the spice of life — John W. Tukey (1980)

MEASUREMENT UNCERTAINTY is the doubt about the true value of the measurand that remains after making a measurement. Measurement uncertainty is described fully and quantitatively by a probability distribution on the set of values of the measurand.

This definition acknowledges that measurement uncertainty is a kind of uncertainty, and decouples the meaning of measurement uncertainty from how it may be represented or described.

The approach is common in mathematics, for example in the definition of function as a set of ordered pairs such that no two different ordered pairs have the same first element.[107] This definition is conceptual and abstract, and quite separate from the consideration of how a function may be evaluated or represented.

[107] R. Dedekind. *The nature and meaning of numbers*. Open Court Publishing Company, Chicago, 1901. Translated from the German by W. W. Beman

Functions can be represented verbally (by stating in plain English what gets mapped to what), using a table that lists the value of the function that corresponds to each value of its argument, graphically, algebraically (by a formula), or algorithmically (describing operations that, once applied to the function's argument, yield the function's value), each enabling a particular way of producing the value of the function that corresponds to a particular value of its argument.

Uncertainty is the absence of certainty, and certainty is either a mental state of belief that is incontrovertible for the holder of the belief (like, "I am certain that my son was born in the month of February"), or a logical necessity (like, "I am certain that 7253 is a prime number").

Uncertainty comes by degrees, and measurement uncertainty, which is a kind of uncertainty, is the degree of separation between a state of knowledge achieved by measurement, and the generally unattainable state of complete and perfect knowledge of the object of measurement.

Measurement uncertainty can be represented most thoroughly by a probability distribution. This representation applies equally well to the measurement of qualitative as of quantitative properties.

When it proves impracticable to express measurement uncertainty quantitatively (either for quantitative or for categorical measurands), it may be expressed using an ordinal scale comprising suitably defined degrees of uncertainty, or levels of confidence. For example, using terms like "virtually certain" or "very likely" in climatology [Mastrandrea et al., 2011]. Similarly, NIST uses words "most confident" and "very confident" to characterize the uncertainty associated with the identity of DNA nucleobases (SRM 2374) or of biological species (SRM 3246 *Ginkgo biloba*).

For quantitative, scalar measurands, measurement uncertainty may be summarily, albeit incompletely, represented by the standard deviation of the corresponding probability distribution, or by similar indications of dispersion. When the chosen summary is the standard deviation, it is usually called the *standard uncertainty*.

Karl Pearson introduced the term "standard deviation" in a lecture that he gave in 1893. George Udny Yule introduced the term "standard error" in 1897 [Yule, 1897], and applied it in the contemporary sense (standard deviation of a function of observations) in 1911 [Yule, 1911].

The *probable error*, which is another indication of dispersion, was far more popular in the 19th and 20th centuries than it is today: for a symmetrical distribution, it is half the length of the interval centered at the center of the distribution that contains 50 % of the distribution's unit of probability. For a Gaussian distribution, the probable error is 67.45 % of the standard deviation, and for a Student's t distribution with 3 degrees of freedom it is 44.16 % of the standard deviation.

Bessel introduced the term "probable error" in the first quarter of the 19th century, and by mid-century its meaning was already well established: "In Astronomy and Physics, when the value of any quantity or element, as the declination of a star, the latitude of a place, the specific gravity of a body, &c., has been determined by means of a number of independent observations, each liable to a small amount of error, the determination (in whatever way it may have been deduced from the observations) will also be liable to some uncertainty; and the *probable error* is the quantity, which is such that there is the same probability of the difference between the determination and the true absolute value of the thing to be determined exceeding or falling short of it" [Brande, 1842, Page 984].

The probable error has an important advantage relative to the standard uncertainty: its meaning is the same regardless of the nature of the underlying (symmetrical) distribution, while the meaning of the standard uncertainty depends markedly on the nature of the distribution, as the illustrations in the probability appendix (Page 159) make abundantly clear.

Another advantage of the probable error is that it exists and is finite for all probability distributions, while some distributions, for example, Student's t distributions (Page 168) with 2 or fewer degrees of freedom, the Cauchy distribution (Page 169) in particular, do not have a finite standard deviation.

In general, a set of selected quantiles (say, the 2.5th, 25th, 50th, 75th, and 97.5th percentiles) of the probability distribution that represents the uncertainty associated with a scalar measurand, provides a more detailed and informative summarization of its dispersion than either the standard uncertainty or the probable error.

The uncertainty surrounding quantitative, multivariate or functional measurands, can be summarized by covariance matrices or by coverage regions, for example coverage bands (Page 102) for calibration and analysis functions (Page 101).

For categorical measurands, the dispersion of the probability distribution over the set of possible values for the property of interest may be summarized by its entropy.[108] Alternatively, the uncertainty may be expressed using rates of false positives and false negatives, sensitivity and specificity,[Altman and Bland, 1994] or receiver operating characteristic curves.[Brown and Davis, 2006]

[108] A. Possolo. *Simple Guide for Evaluating and Expressing the Uncertainty of NIST Measurement Results.* National Institute of Standards and Technology, Gaithersburg, MD, 2015. doi:10.6028/NIST.TN.1900. NIST Technical Note 1900

The GUM distinguishes two manners of evaluating the contributions that different sources of uncertainty make to the uncertainty associated with an estimate of a measurand, and calls them *Type A* and *Type B*. The former involves application of a "method of evaluation of uncertainty by the statistical analysis of series of observations" (GUM 2.3.2), and the latter refers to any other method.

The names *Type A* and *Type B* were chosen because those involved in laying the foundations for what became the GUM, during the 1980s, were unable to agree on more descriptive terms (Ronald Collé, 2012, *personal comm.*).

However, about fifteen years earlier, Mosteller and Tukey [1986] had already proposed perfectly serviceable alternatives: "We assign contributions to uncertainty to two sources: those that might be judged from the data at hand — internal uncertainty; and those that come from causes whose effects are not revealed by the data — supplementary uncertainty."

Appendix: Probability

Imagine an explorer looking for the wreckage of an airplane resting at the bottom of the sea, with the aid of a map where shades of gray represent probabilities for the location of the wreckage. A *probability distribution* is like this map, or like a distribution of mass over the set of possible values for a measurand: where the shades of gray are darkest, or where the mass density is largest, is where the true value of the measurand most likely lies.

Probability map for the location of Air France 447 site after three unsuccessful searches from June 2009 to May 2010. Darker areas indicate highest probability for the wreckage's location, and the white cross shows the location of where the wreckage was found in 2011. Modified version of Figure 33 in Stone et al. [2011].

PROBABILITY DISTRIBUTIONS over sets of values of quantities or qualities are mathematical objects very similar to distributions of mass in space. Probability, the same as mass, may be distributed continuously, smoothly (as one spreads jelly on bread), or it may be distributed discretely, in lumps (as one places blobs of cookie dough on a baking sheet).

A distribution of probability, like a distribution of mass, may have both features: being smooth in some regions, and lumpy in others. For example, an estimate of dark uncertainty (discussed under *Consensus Building*, Page 146) typically can be zero with positive probability, hence its probability distribution places a lump of probability at 0, but the rest of the probability is distributed smoothly over all positive numbers.

The set to which a probability distribution assigns its unit of probability is called the *support* of the distribution. But while masses of all sizes can be distributed over regions of space, all probability distributions have available a single unit of probability to spread around. Where probability piles up and stands out it suggests where it is more likely that the treasure lies.

The same as with mass, if a probability distribution spreads its unit of probability continuously and smoothly over the set of values where it is defined, then one can speak of its *probability density*. If the probability distribution spreads its unit in discrete lumps over a denumerable set of values, then the function that produces the probability in each lump is called its *probability mass function.*

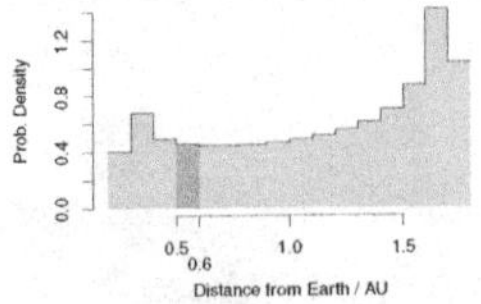

Histogram depicting the probability of finding Venus within particular distances from Earth. The lightly shaded rectangle from 0.5 AU to 0.6 AU, has an area 0.04544 which is the probability that, on a randomly chosen day, Venus will be between 0.5 AU and 0.6 AU from Earth. This probability was computed by determining the number of days, between December 25th, 2020, and December 24th, 2420, when the distance to Venus will be in that interval, and dividing it by the total number of days in this period: 6638/146097 = 0.04544.

Consider the probability density of the distance from Earth to Venus as both planets travel in their orbits around the Sun. The function whose graph is the black polygonal line that tops the histogram depicted alongside, is a *probability density function*: it represents probabilities by areas under its graph. The total area shaded light or dark gray is 1.

The assignment of the unit of probability to the horizontal axis according to the areas under the polygonal line defines what is called a *probability distribution* on this axis. In this case, probability piles up toward the ends of the range of distances, and it is scarcer in the middle.

If the area under the polygonal line is conceived as representing matter of uniform density, and this matter collapses to form a rigid, cylindrical rod on the horizontal axis, then the probability distribution is the distribution of mass of this rod, and the probability density function depicts the variation of the mass density along the rod.

Suppose that the function P defines the probability distribution of the distance, D, from Venus to Earth so that $P(I)$ denotes the probability of this distance having a value in a specified interval $I = (a, b)$. If P is such that $P(I) = \int_a^b p(s)(d)s$ for some non-negative function p, then we say that p is the probability density function of P. In these circumstances, the mean of the probability distribution is $\mu = \int_{\mathcal{S}} sp(s)(d)s$, and its variance is $\sigma^2 = \int_{\mathcal{S}} (s-\mu)^2 p(s)(d)s$, where $\mathcal{S}$ denotes the set of values of D such that $P(\mathcal{S}) = 1$.

The mean of the distribution is the rod's center of mass, and the variance is the rod's second moment of inertia when the rod rotates about its center of mass, with axis of rotation perpendicular to the rod.

Probability distributions naturally arrange themselves into families: Gaussian distributions, Weibull distributions, etc. The members of the same family have probability densities of the same form, differing only in the values of some *parameters,* which identify the individual members of the family. For example, individual Gaussian distributions are identified by the mean and standard deviation, and individual Weibull distributions by the shape and scale parameters.

Odds are often used to express probabilities. For example, the morning line odds posted for *Tiz The Law* to win the 2020 Kentucky Derby, were 3 to 5 (often written as 3:5, meaning $3/5 = 0.6$). This ratio, times the amount bet, is the prize earned in case of a win: had *Tiz The Law* won the race, a \$2 bet on it would have earned a prize of $(3/5) \times \$2 = \$1.20$, hence a payout of \$3.20. But the winner of the 2020 Kentucky Derby was *Authentic*, whose odds had been 8 to 1.

In principle, those odds should have meant that the probability of *Tiz The Law* winning the race was $5/(3+5) = 62.5\,\%$, because this would have made the gamble fair: $(5/8) \times \$1.20 - (3/8) \times \$2 = 0$. However, real-life gambles are never fair: the sum of the implied probabilities corresponding to the morning line odds for the 18 horses set to Run for the Roses was 1.35, thus forming a *Dutch Book*.[109]

[109] A. Hájek. Dutch Book Arguments. In P. Anand, P. K. Pattanaik, and C. Puppe, editors, *The Handbook of Rational & Social Choice*, chapter 7, pages 173–195. Oxford University Press, Oxford, UK, 2009. ISBN 978-0-19-929042-0

Betting odds generally are odds *against*. In general, the relation between the odds o (in favor) of an event, and the event's probability p are

$$o = p/(1-p), \quad \text{or} \quad p = o/(1+o).$$

Random variables are quantities or qualities that have a probability distribution as an attribute. This attribute serves to indicate which subsets of their respective ranges (the sets where they take their values) are more likely to contain the value that the random variable takes when it is *realized*.

Volnay Clos des Chênes (Domaine Michel Lafarge), from the Côte-d'Or, France.

For example, the volume of wine in a bottle of *Volnay Clos des Chênes (Domaine Michel Lafarge)*, from the Côte-d'Or, France, is a (quantitative) random variable that is realized every time a bottle is filled at the winery. The probability distribution of this random variable is continuous, and it is concentrated in a narrow range around 750 mL.

The identity of the nucleotide at a particular locus of a strand of DNA is a (qualitative) random variable whose possible values are adenine, cytosine, guanine, and thymine, and whose realized value is the identity of the nucleotide that is actually there. The probability distribution of this random variable is discrete, its unit of probability being allocated in lumps to those four possible compounds: for example, the probability is 30 % of finding adenine at any particular locus in the human genome.

The adjective *random* in the expression "random variable X" bears no metaphysical connotation: in particular, it does not suggest that X is an outcome of a game of chance that Nature is playing against us. It is merely a mnemonic and allusive device to remind us that X has a probability distribution as an attribute.

Suppose the random variable in question is the Newtonian constant of gravitation, G, which is generally believed to be constant, but whose true value is known only imperfectly. A probability distribution with a relative standard deviation that currently stands at 2.2×10^{-5} can be used to describe the corresponding uncertainty [Tiesinga et al., 2021].

Similarly, the distance between Venus and Earth can also be characterized as a random variable, even if its value is predictable deterministically as a function of the date. The probability distribution of this random variable can describe the uncertainty associated with that distance on a day chosen at random.

The probability distribution of a random variable determines the probability that it will take a value in any given subset of its range. The corresponding computation is particularly easy when the probability distribution has a probability density (Page 159) that is specified analytically. How this is done depends on whether the distribution of the random variable is continuous, discrete, or of a mixed type (that is, has a continuous component over its range, as well as lumps of probability at some of the values in its range).

Suppose that X is a scalar random variable (for example, the lifetime of a 25 W incandescent light bulb GE A19, whose expected lifetime is 2000 h) and that its probability distribution is continuous and has probability density p_X. Then, the probability that X takes a value in a set A (which may be an interval or a more complicated set), and which we write as $\Pr\{X \in A\}$, is the area under the graph of p_X over the set A. If X has an exponential probability distribution, with density $p_X(x) = \lambda \exp(-\lambda x)$ and mean $\lambda^{-1} = 2000\,\text{h}$ (as depicted alongside), then $\Pr\{3000\,\text{h} < X < 4000\,\text{h}\}$ is the shaded area, which in this case can be computed analytically:

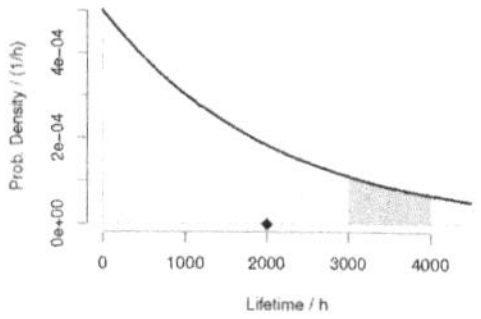

Probability density (Page 159) of the lifetime of a GE A19 25 W incandescent light bulb. The small diamond marks its expected lifetime, and the shaded area is the probability that the bulb will last between 3000 h and 4000 h.

$$\int_{3000}^{4000} \frac{1}{2000} \exp(-x/2000)\,\mathrm{d}x = 0.09.$$

INDEPENDENCE is an important property and a consequential assumption. Two random variables, X and Y, are said to be *independent* when the probability that X takes a value in a set A, and that Y takes a value in a set B, when both are realized jointly, or simultaneously, is equal to the product of their individual probabilities of taking values in such sets one separately from the other.

The uncertainty of the average of replicated, independent determinations of the same quantity generally will be smaller than the uncertainty of any individual measurement — the prize of claiming independence.

Consider three such determinations with the same standard uncertainty. If modeled as outcomes of independent random variables, then their average will have a standard uncertainty that is $\sqrt{3}$ times smaller than the standard uncertainty of the individual determinations. If, however, they all are affected by the same error (for example, resulting from miscalibration of the measuring instrument used to obtain them), then averaging the replicates will not reduce the uncertainty component attributable to miscalibration.

For example, the number of Aces in a poker hand, and the number of cards from the suit of diamonds in the same hand, are dependent random variables, because knowing that there are five diamonds implies that there cannot be more than one Ace.

Independence is next to impossible to verify empirically in most cases, because doing so involves showing that $\Pr(X{\in}A \text{ and } Y{\in}B) = \Pr(X{\in}A) \times \Pr(Y{\in}B)$ for all subsets A and B of the respective ranges of X and Y. If these ranges have infinitely many values, then this verification requires an infinitely large experiment.

Two events are independent when the probability of

their joint occurrence is equal to the product of their individual probabilities. If the probability of one of them occurring depends on the knowledge of the other one having occurred or not, then the events are dependent.

Consider rolling two casino dice (perfectly shaped and balanced cubes with 1, 2, ..., 6 pips on their faces), one red and the other blue, and the following two events: A is getting 3 pips on the red die; B is getting 7 pips in total. The probability of A is 1/6, the probability of B is 6/36, and the probability of A and B both occurring simultaneously is $\Pr(A \text{ and } B) = 1/36 = (1/6) \times (6/36) = \Pr(A) \times \Pr(B)$: therefore, A and B are independent events. However, getting 3 pips on the red die, and 8 pips in total, are dependent events.

When one says that $\{x_1, \dots, x_n\}$ is a *sample* from a probability distribution, one means that these are outcomes of n independent, identically distributed random variables whose common distribution is the distribution that the sample allegedly comes from.

EXCHANGEABLE RANDOM VARIABLES are such that the random vectors $(X_1, \dots, X_n)$ and $(X_{\pi(1)}, \dots, X_{\pi(n)})$ have the same joint probability distribution, for any permutation π of the indices $\{1, \dots, n\}$. Exchangeable random variables have identical (marginal) distributions, but generally they are dependent, with correlations never smaller than $-1/(n-1)$.

If a standard deck of cards used for playing poker is well-shuffled, then the numbers of aces in the hands dealt to the players are exchangeable but dependent random variables.

Zabell [2005, Chapter 4] attributes the origin of the concept to the English philosopher W. E. Johnson,[110] even if it was popularized and developed by Bruno de Finetti.[111]

[110] W. E. Johnson. *Logic, Part III — The Logical Foundations of Science*. Cambridge University Press, London, UK, 1924

[111] B. de Finetti. Funzione caratteristica di un fenomeno aleatorio. *Atti della R. Academia Nazionale dei Lincei, Serie 6. Memorie, Classe di Scienze Fisiche, Mathematice e Naturale*, 4:251–299, 1930

Exchangeability is often established via symmetry arguments. For example, when considering a set of triplicate determinations, $\{w_1, w_2, w_3\}$, of the mass fraction of arsenic in a unit of NIST SRM 3268 *Pueraria montana* var. *lobata* (Kudzu) extract (Page 151), we may conclude that the order in which the determinations were made is irrelevant for any conclusions to be derived from them, hence that they are exchangeable.

But this then implies that there is a *prior* probability distribution for their true mean, θ, whose density (Page 159) p is such that the density g of the joint probability distribution of those replicates is of the form

$$g(w_1, w_2, w_3) = \int_0^{+\infty} f(w_1|\theta) f(w_2|\theta) f(w_3|\theta) p(\theta) \mathrm{d}\theta,$$

where f denotes the probability density of the sampling distribution of the observed mass fractions [Bernardo, 1996]. Thus, the assumption of exchangeability naturally suggests a Bayesian treatment of the determinations of the mass fraction of arsenic in kudzu.

Median, Mean, Variance, Bias, and Mean Squared Error are properties of random variables (or of their probability distributions). The median is meaningful for random variables whose values are ordinal or quantitative and scalar, while the mean, variance, bias, and mean squared error are meaningful only for quantitative, scalar properties.

Median, mean, and variance are intrinsic properties of the random variables, while bias and mean squared error become meaningful when a random variable plays the role of estimator of a quantity whose true value is unknown.

The median of a random variable is any value such that the random variable is equally likely to take values smaller or larger than it. The mean of a random variable is the center of mass of its probability distribution, when the distribution is regarded as the distribution of a unit of mass over the range of the random variable. And its variance is the second moment of inertia of such distribution of mass, about its mean.

The mean is also called the *expected value* (mathematical expectation), and for this reason the mean of the random variable X is often denoted $\mathbb{E}(X)$. The variance is the expected value of the squared difference between a random variable and its mean,

$$\mathbb{V}(X) = \mathbb{E}[X - \mathbb{E}(X)]^2,$$

and the standard deviation is the (positive) square root of the variance: it describes how scattered around the mean the unit of probability is.

A random variable can be so unpredictable that its mean does not exist (this is the case for the reciprocal of a Gaussian random variable), or its variance is infinite. However, the median of any random variable X, $\mathbb{M}(X)$ (this notation for the median is not standard), always exists and is finite, and so is another indication of how dispersed the corresponding probability distribution is, the median absolute deviation from the median, $\mathbb{M}(X - \mathbb{M}(X))$, often abbreviated as MAD.

If X has a discrete distribution, and the different values that it can take are $x_1, x_2, \ldots$, then

$$\mathbb{E}(X) = x_1 p_1 + x_2 p_2 + \ldots,$$

where $p_i = \Pr(X = x_i)$ for $i = 1, 2, \ldots$, provided this sum, which may involve infinitely many summands, is finite. If X has a continuous distribution with probability density (Page 159) p, then

$$\mathbb{E}(X) = \int_{\mathcal{X}} x p(x) \mathrm{d}x,$$

where $\mathcal{X}$ denotes the range of X, provided this integral converges.

Now suppose that a random variable X is to play the role of estimator of a quantity θ whose value is unknown. For example, X may be the mass fraction of inorganic arsenic in a sample of shellfish tissue, and θ may be the true mass fraction of arsenic in it. Owing to incomplete extraction of the arsenic during sample preparation, the expected value of X may well be less than θ.

The *bias* of X as estimator of θ is the difference between its expected and true values, $\mathbb{E}(X) - \theta$. The *mean squared error* (MSE) of X as estimator of θ is the bias squared plus the variance, $(\mathbb{E}(X) - \theta)^2 + \mathbb{V}(X)$.

If X and Y are scalar random variables, and a and b are real numbers, then $\mathbb{E}(aX + bY) = a\mathbb{E}(X) + b\mathbb{E}(Y)$, regardless of whether X and Y are dependent or independent. And if X and Y are independent, then $\mathbb{E}(XY) = \mathbb{E}(X) \times \mathbb{E}(Y)$, and $\mathbb{V}(aX + bY) = a^2\mathbb{V}(X) + b^2\mathbb{V}(Y)$. In particular, note that $\mathbb{V}(X - Y) = \mathbb{V}(X + Y) = \mathbb{V}(X) + \mathbb{V}(Y)$, provided X and Y are independent.

The uniform (or rectangular) probability distribution over an interval $[a, b]$, where $a < b$, is a continuous distribution whose probability density function is constant and equal to $1/(b - a)$ over that interval, and zero everywhere else. Since its graph is a rectangle, the distribution is also called *rectangular*.

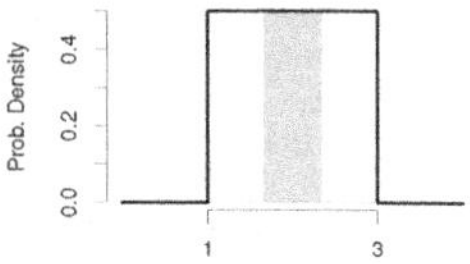

Probability density (Page 159) of the uniform distribution on the interval $[1, 3]$, with mean 2 and standard deviation $1/\sqrt{3} = 0.58$. The shaded region comprises 68 % of the area under the curve.

This distribution has mean $\mu = (a + b)/2$ and standard deviation $\sigma = (b - a)/\sqrt{12}$. Since probabilities are given by areas under the graph of the probability density, the probability that a uniform distribution assigns to an interval $[x - \delta, x + \delta]$, for some $\delta > 0$ and any real number x, either is zero or decreases to zero as δ decreases to zero.

The uniform distribution, and indeed every continuous distribution, thus has the apparently paradoxical property that even though it assigns probability zero to every individual real number, the probability it assigns to all of them together still adds up to 1.

The Gaussian (or normal) probability distribution with mean μ and standard deviation $\sigma > 0$ is a continuous distribution on the infinitely long interval that comprises all real numbers. Its probability den-

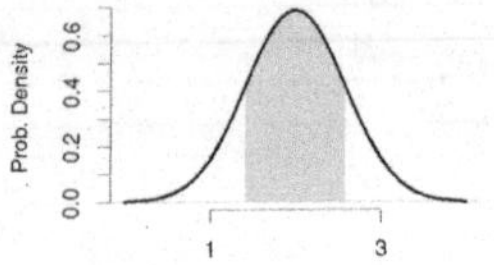

Probability density of the Gaussian distribution with mean 2 and standard deviation $1/\sqrt{3} = 0.58$. The shaded region comprises 68 % of the area under the curve.

sity (Page 159) has the familiar bell-shaped curve as its graph: it is symmetrical around μ and has inflection points at $\mu \pm \sigma$. The area under the curve between the inflection points is 68 %, and the corresponding area between $\mu \pm 2\sigma$ is 95 % approximately.

The Gaussian distribution plays a central role in probability theory because the probability distribution of the sum of several independent random variables can, under very general conditions, be approximated by a Gaussian distribution — a remarkable fact first established in fair generality by Pierre Simon, Marquis de Laplace, in 1812.

A unique, surprising property of the Gaussian distribution is that "a necessary and sufficient condition for the normality of the parent distribution is that the sampling distributions of the mean and of the variance be independent."[112] This is surprising because both the sample average and the sample variance are functions of the same data.

[112] E. Lukacs. A characterization of the normal distribution. *Annals of Mathematical Statistics*, 13(1):91–93, March 1942. doi:10.1214/aoms/1177731647

The distribution takes its name from Carl Friedrich Gauss (1777–1855) because he proved that the arithmetic average is the best combination of observations (in the sense of minimizing mean squared error) when the errors of observation are Gaussian, thus providing a rationale for the widespread practice of averaging observations.

"The reference standard for shapes of distribution has long been the shape associated with the name of Gauss, who combined mathematical genius with great experience with the highest-quality data of his day — that of surveying and astronomy. Later writers have made the mistake of thinking that the Gaussian (sometimes misleadingly called normal) distribution was a physical law to which data must adhere — rather than a reference standard against which its discrepancies are to be made plain." — John W. Tukey (1977, §19B)

The distribution is also called "normal." However, John Tukey in particular, has made clear that it is far from being a universally adequate model for data. On the contrary, he places the Gaussian distribution among the defining elements of what he calls the *utopian* situation for data analysis — an "ideal" situation that is as mathematically convenient as it often is disjointed from reality.

The Student's t probability distribution is determined by its median (which can be positive, negative, or zero), scale (which must be positive), and a positive (but not necessarily integer) number of degrees of freedom $\nu > 0$. It is a continuous distribution on the set of all real numbers.

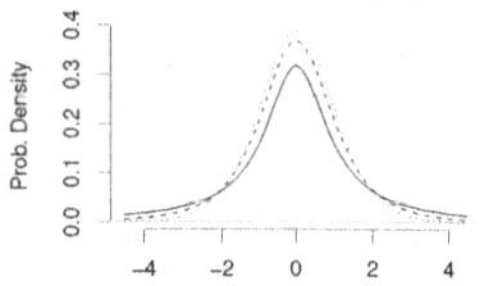

Probability densities of Student's t distributions with center 0, scale 1, and number of degrees of freedom 1 (solid line), 3 (dashed line), and 9 (dotted line).

The graph of Student's probability density (Page 159) also is bell shaped, but its tails are heavier (and its center lighter) than in the Gaussian distribution with the same mean and standard deviation. The parameter ν controls its tail heaviness: the smaller the ν, the heavier the tails. For example, Student's t distribution with mean 0 and standard deviation $\sqrt{3}$ (which has 3 degrees of freedom), assigns almost seven times more probability to the interval $[6, 7]$ than a Gaussian distribution with the same mean and standard deviation.

This distribution is remarkable, and pervasive, owing to this fact: if $x_1, \ldots, x_m$ are a sample of size m drawn from a Gaussian distribution with mean μ and standard deviation σ, $\overline{x}_m$ is the sample average, and

$$s_m^2 = \frac{(x_1 - \overline{x}_m)^2 + \cdots + (x_m - \overline{x}_m)^2}{m - 1}$$

is the sample variance, then

$$\frac{\overline{x}_m - \mu}{s_m / \sqrt{m}}$$

is like an outcome of a random variable with a Student's t distribution with center 0, scale 1, and $m-1$ degrees of freedom. Remarkably, the probability distribution of this ratio does not involve the unknown σ.

If $\nu \leqslant 2$, then the Student's t distribution has infinite variance. A Student's t distribution with $\nu = 1$ is called a Cauchy or Lorentz distribution: it has neither variance nor mean. Random variables with Cauchy distributions are truly wild things. This is how wild: the average of a sample from a Cauchy distribution has the same distribution as the individual sample values.

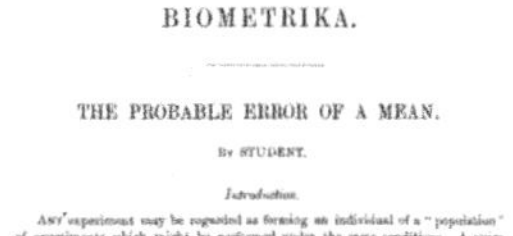
BIOMETRIKA.

THE PROBABLE ERROR OF A MEAN.

By STUDENT.

Introduction.

Any experiment may be regarded as forming an individual of a "population" of experiments which might be performed under the same conditions. A series of experiments is a sample drawn from this population.

This distribution is called "Student's" because statistician William Sealy Gosset (1876–1937) published an article introducing the use of this distribution under the pseudonym "Student" [Student, 1908].

Gosset was employed by the *Guinness* brewery in Dublin, and legend holds that his use of a pseudonym was due to the company's concern for secrecy in their use of statistical methods for quality control [Wendl, 2016].

Zabell [2008] reminds us that in 1930 Harold Hotelling noted that this anonymity was designed to hide Gosset's identity not from the outside world but from his own colleagues at *Guinness*.

The Half-Cauchy probability distribution is a continuous distribution that results from "folding" at zero a Cauchy distribution centered at zero, so that the 50 % probability it assigns to the negative numbers is transferred to the positive numbers as by a mirror placed at zero. Gelman [2006] suggests the half-Cauchy as a general purpose, weakly informative prior distribution for standard deviations in Bayesian random effects models. We use it in this role when computing a consensus value for the mass fraction of arsenic in kudzu (Page 151).

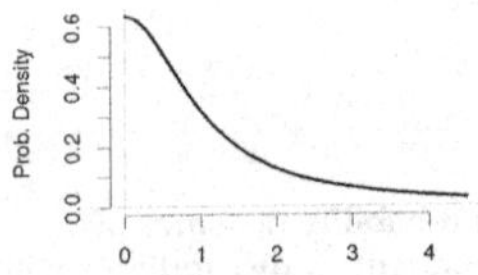

Probability density (Page 159) of the Half-Cauchy distribution with median 1.

The Gamma probability distribution is a continuous distribution determined by two positive parameters, shape α and rate λ, concentrated on the positive real numbers. The distribution is skewed to the right, with a right tail longer than the left tail. The mean of the gamma distribution is α/λ, and the variance is α/λ^2. The **chi-square probability distribution** with ν degrees of freedom is a gamma distribution with shape $\alpha = \nu/2$, and rate $\lambda = ½$, hence its mean is ν and its variance is 2ν.

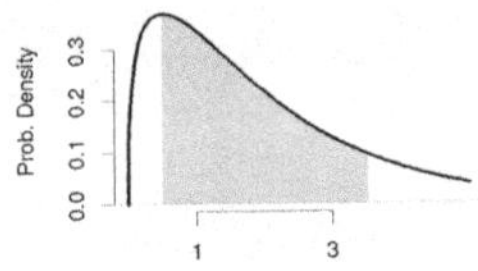

Probability density (Page 159) of the gamma distribution with mean 2 and standard deviation $1/\sqrt{3}$. The shaded region comprises 68 % of the area under the curve.

A gamma distribution with shape $\alpha = 1.7$ and rate $\lambda = 762\,\text{kg/mg}$ is used in the measurement of nitrite in seawater (Page 214) to encapsulate prior knowledge about measurement uncertainty associated with Griess's method.

The steadily decreasing curve is the probability density (Page 159) of the chi-square distribution with both mean and standard deviation equal to 2. The shaded region comprises 68 % of the area under the curve. When the number of degrees of freedom ν is greater than 2, the curve has a single hump, reaching a maximum at $\nu - 2$.

The Gaussian, chi-square, and Student's t distributions are related in a remarkable manner. If $\overline{x}$ and s are the average and standard deviation of a sample of size m drawn from a Gaussian distribution whose mean μ and standard deviation σ both are unknown, then: (i) $\overline{x}$ and s are like outcomes of two independent random variables (even though they are functions of the same data); (ii) $(m-1)s^2/\sigma^2$ is like an outcome of a chi-square random variable with $m-1$ degrees of freedom; and (iii) $(\overline{x}-\mu)/(s/\sqrt{m})$ is like an outcome of a Student's t

random variable with $n-1$ degrees of freedom, hence its distribution does not depend on the unknown σ. This last fact is the basis for the coverage intervals specified in Annex G of the GUM [JCGM 100:2008].

The Stan code that implements a random effects model for the determinations of the mass fraction of arsenic in kudzu employs the chi-square distribution in the likelihood function to express the uncertainty associated with sample standard deviations based on small numbers of degrees of freedom as follows: if ν denotes the number of degrees of freedom that s is based on, then $\nu s^2/\sigma^2$ is like an outcome of a chi-square random variable with ν degrees of freedom, and s^2 is like an outcome of a gamma random variable with shape $\nu/2$ and rate $\nu/(2\sigma^2)$.

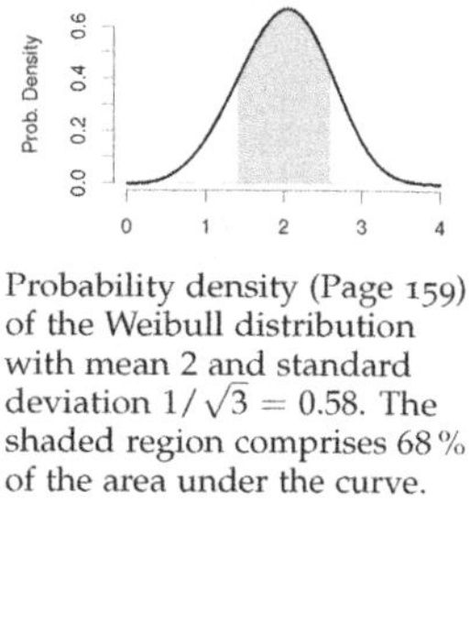

Probability density (Page 159) of the Weibull distribution with mean 2 and standard deviation $1/\sqrt{3} = 0.58$. The shaded region comprises 68 % of the area under the curve.

THE WEIBULL PROBABILITY DISTRIBUTION may be the most important continuous, univariate distribution, after the Gaussian, chi-square, and Student's t distributions. The Weibull distribution is concentrated on the positive real numbers, and it is indexed by two parameters: shape $\alpha > 0$ and scale $\eta > 0$, with mean $\eta\Gamma(1+1/\alpha)$ and standard deviation $\eta(\Gamma(1+2/\alpha) - \Gamma^2(1+1/\alpha))^{1/2}$, where "$\Gamma$" denotes the gamma function of mathematical analysis (whose values can be computed in R using function `gamma`).

The Weibull distribution is renowned for being an accurate model for the strength of many materials, and for the longevity of mechanical parts and machinery. Its parameters can easily be estimated from failure data by application of either the method of maximum likelihood (Page 193) or Bayes methods (Page 208).

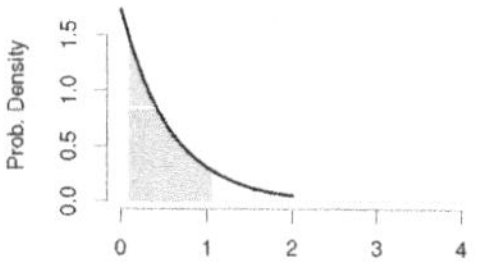

Probability density (Page 159) of the exponential distribution with mean and standard deviation both equal to $1/\sqrt{3} = 0.58$. The shaded region comprises 68 % of the area under the curve.

The exponential distribution with rate $\lambda > 0$ is a Weibull distribution with shape 1 and scale $1/\lambda$, hence with mean and standard deviation both equal to $1/\lambda$, and median $\ln(2)/\lambda$. The exponential distribution is a suitable model for the lifetime of an item that does not age with passing time (for example, individual atoms of ^{222}Rn, Page 46): the probability that an exponential lifetime will last longer than $t + \Delta$, given that it has lasted t already, is the same as the probability that it will have lasted longer than Δ to begin with. The shortest of several independent, exponentially distributed lifetimes is exponentially distributed with rate equal to the sum of all their rates.

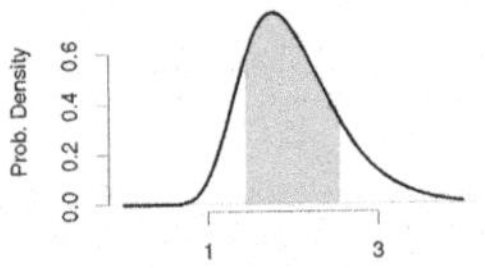

Probability density (Page 159) of the lognormal distribution with mean 2 and standard deviation $1/\sqrt{3} = 0.58$. The shaded region comprises 68 % of the area under the curve.

THE LOGNORMAL PROBABILITY DISTRIBUTION is a continuous distribution concentrated on the positive real numbers. If a random variable X has a lognormal distribution with mean μ and standard deviation $\sigma > 0$, then $\ln(X)$ has a Gaussian distribution with mean $\ln(\mu/\sqrt{(\sigma/\mu)^2+1})$, and variance $\ln((\sigma/\mu)^2 + 1)$.

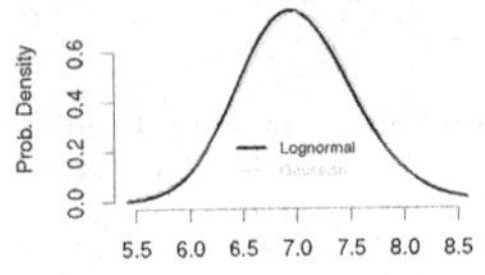

Probability densities of the lognormal (black thick curve) and Gaussian (gray thin curve) distributions, both with mean 7 and standard deviation 0.525. The coefficient of variation is 7.5 %, and the two densities already provide a close approximation to one another.

Ratios, U/V, arise often in metrology, and the Gaussian distribution just as often is the natural candidate to model the uncertainties that surround them. However, assigning a Gaussian distribution to the denominator, V, implies that the probability is positive that V shall take a value arbitrarily close to zero, hence that the ratio may become arbitrarily large in absolute value, or, in other words, that the uncertainty of the ratio will be infinite. Of course, if zero lies many standard deviations away from V's expected value, then this difficulty may not matter in practice.

When the coefficient of variation of V (standard deviation divided by the mean) is small (less than 5 %), then Gaussian and lognormal distributions with identical means and with identical standard deviations will be essentially identical, and the lognormal model may be used to avoid the possibility of inducing an unrealistically large variance for the ratio.

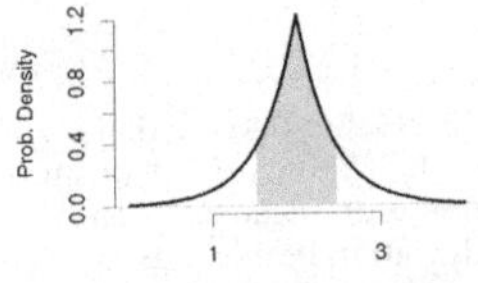

Probability density (Page 159) of the Laplace distribution with mean 2 and standard deviation $1/\sqrt{3} = 0.58$. The shaded region comprises 68 % of the area under the curve.

THE LAPLACE PROBABILITY DISTRIBUTION, also called the double exponential distribution, is a continuous distribution specified by its mean and scale parameters. Its standard deviation is $\sqrt{2}$ larger than the scale parameter.

We use the Laplace distribution in a model for the results of an interlaboratory study of the mass fraction of arsenic in kudzu (Page 151) because its tails are heavier than the tails of the Gaussian distribution with the same mean and standard deviation, thus reducing the influence that measured values far from the bulk of the others have upon the consensus value.

THE BINOMIAL PROBABILITY DISTRIBUTION is a discrete distribution that assigns lumps of probability to the non-negative integers $0, 1, \ldots, n$ for some integer $n > 0$. It describes the variability of the number of "successes" in n independent trials whose outcomes are either a "success" or a "failure," when each trial has the same probability $0 \leqslant \theta \leqslant 1$ of yielding a "success." It has mean $n\theta$ and standard deviation $\sqrt{n\theta(1-\theta)}$.

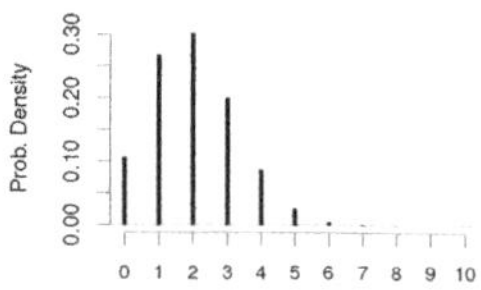

Probability mass function of the Binomial distribution with mean 2, based on 10 trials, with 0.2 probability of "success" in each trial.

The binomial distribution is often used to characterize the uncertainty surrounding the number of entities of a particular type that are being identified and counted in a collection of similar entities. For example, of the number of eosinophils in a collection of 100 white blood cells that are being identified and counted in a differential leukocyte count (Page 38). This count may be regarded as outcome of a binomial random variable based on 100 *trials* (examinations of individual cells), each of which yields a eosinophil (*success*) or a white blood cell of some other type (*failure*).

THE POISSON PROBABILITY DISTRIBUTION is a discrete distribution concentrated on the non-negative integers, whose mean and variance are identical. The probability that a Poisson random variable with mean $\lambda > 0$ will take the value x is $e^{-\lambda}\lambda^x/x!$, where $x! = x(x-1)(x-2)\ldots 1$.

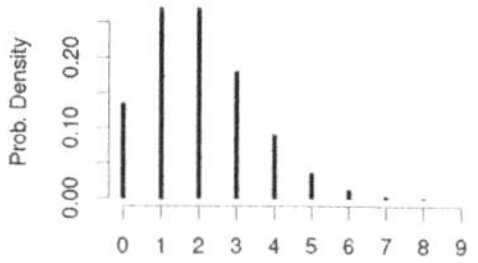

Probability mass function of the Poisson distribution with mean 2.

The number of alpha particles emitted per second and per nanogram of ^{226}Ra, as a result of radioactive disintegration, is a Poisson random variable with mean

$\lambda = 36.6\,\mathrm{s}^{-1}\,\mathrm{ng}^{-1}$. Starting with a particular amount of a radioisotope with a single mode of decay, for example ^{226}Ra, the expected number of decays per second decreases (exponentially fast) over time because one atom of the radioisotope is lost with each decay.

Considering that no two different atoms decay simultaneously, and that the numbers of decays occurring during non-overlapping time intervals are independent (but not identically distributed) random variables, the sequence of epochs at which a decay occurs is a non-stationary (more commonly called *inhomogeneous*) Poisson process.[113]

[113] A. F. Karr. Poisson process. In S. Kotz, C. B. Read, N. Balakrishnan, B. Vidakovic, and N. L. Johnson, editors, *Encyclopedia of Statistical Sciences*, volume 3, pages 1910–1918. John Wiley & Sons, Hoboken, NJ, second edition, 2006. ISBN 978-0-471-15044-2. doi:10.1002/0471667196

This implies that the number of decays occurring during the time interval (t_1, t_2) is a Poisson random variable with mean $\int_{t_1}^{t_2} \lambda(t)\mathrm{d}t$ regardless of the duration of that interval relative to the half-life of the radioisotope, where $\lambda(t)$ denotes the intensity (instantaneous mean number of decays per second) of the Poisson process at time t.

Alleged limitations of Poisson statistics in describing radioactive decay are a misunderstanding caused by the failure to recognize that the underlying Poisson process is inhomogeneous.[114]

[114] A. Sitek and A. M. Celler. Limitations of Poisson statistics in describing radioactive decay. *Physica Medica*, 31:1105–1107, 2015. doi:10.1016/j.ejmp.2015.08.015

This can be taken into account by focusing on the number of decays per unit of time and per mole of the radioisotope at each instant in time, which is equivalent to rescaling the axis of time so that the Poisson process becomes homogeneous [Snyder and Miller, 1991, Problem 2.3.5].

Alternatively, the decreasing intensity of the process can also be dealt with by partitioning the time interval of interest into a set of sufficiently short sub-intervals, and considering a sum of independent, binomial random variables whose different probabilities of "success" are the Poisson probabilities of decay in these sub-intervals.

Of course, if the interval of interest is short by comparison with the radionuclide's half-life, then treating the inhomogeneous Poisson process as if it were homogeneous provides yet another approximate solution.

The numbers of boys (Page 141) that were sick in bed on each day of an influenza epidemic in an English boarding school were modeled as outcomes of independent, Poisson random variables whose means varied from day to day.

However, this was only a convenient approximation because each daily count of sick boys is a sum of dependent Bernoulli random variables, owing to influenza being a contagious disease. (A Bernoulli random variable has a binomial distribution (Page 173) with $n = 1$ trial, hence it can take only the values 0 or 1.)

The Poisson distribution is often used as a model for the number of occurrences of a rare event because Poisson probabilities can approximate binomial probabilities (Page 173) quite closely, when the probability of "success" is small.

A river's 100-year flood (Page 182) is a rare event whose probability of occurrence on any particular year is, by definition, 0.01. The binomial probability (Page 173) of it occurring exactly once (meaning once and once only) in a century is $100(0.01)^1(1-0.01)^{99} = 0.3697$. The corresponding Poisson approximation is computed by putting $x = 1$ and $\lambda = 100 \times 0.01 = 1$ in the formula above, to get $e^{-1}(1)^1/1! = 0.3679$.

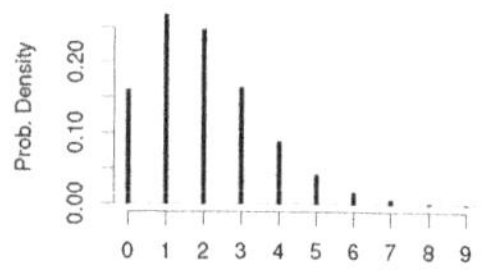

Probability mass function of the Negative Binomial distribution with mean 2 and dispersion $\phi = 10$. The smaller the ϕ the greater the dispersion.

THE NEGATIVE BINOMIAL PROBABILITY DISTRIBUTION with mean $\mu > 0$ and dispersion $\phi > 0$ is a discrete distribution concentrated on the non-negative integers $0, 1, 2, \ldots$. Its variance is $\mu + \mu^2/\phi$, hence it is larger than the variance of a Poisson distribution with the same mean. For this reason it is often used as a model for counts that are more dispersed than Poisson counts.

THE MULTINOMIAL PROBABILITY DISTRIBUTION assigns its unit of probability to K different sets or *categories*, so that set $k = 1, \ldots, K$ receives probability $\theta_k \geqslant 0$ and $\theta_1 + \cdots + \theta_K = 1$. Identifying and counting 100 leukocytes is equivalent to placing 100 balls into 7 bins, the balls representing leukocytes and the bins representing the types of leukocytes. The probabilities $\{\theta_k\}$ may be estimated by the relative frequencies of the different types of leukocytes. In general, if n denotes the number of items to be categorized and counted, then the mean number of items expected for category k is $n\theta_k$, and the standard deviation of this number is $n\theta_k(1 - \theta_k)$. The correlation between the numbers of items in categories $1 \leqslant j < k \leqslant K$ is $-\sqrt{\theta_j\theta_k/((1-\theta_j)(1-\theta_k))}$. Note that all the correlations are negative because an overcount in one category will induce an undercount in another.

Appendix: Statistics

The great American statistician Jimmy Savage defined "statistics proper" as "the art of dealing with vagueness and with interpersonal difference in decision situations." [Savage, 1972, Chapter 8] The focus on decision-making suggests an action oriented discipline, "vagueness" refers to uncertainty, whereas the interpersonal difference comprises all differences of taste and differences of judgment, both typically varying from person to person.

"The evaluation of uncertainty is neither a routine task nor a purely mathematical one; it depends on detailed knowledge of the nature of the measurand and of the measurement." — GUM 3.4.8 [JCGM 100:2008].

Similarly to how "It is only slightly overstating the case to say that physics is the study of symmetry,"[115] one can even perhaps say that statistics is the alchemy of distilling uncertainty into some certainty, in the sense that "the history of data analysis can be read as a succession of searches for certainty about uncertainty".[116]

[115] P. W. Anderson. More is different: Broken symmetry and the nature of the hierarchical structure of science. *Science*, 177:393–396, 1972. doi:10.1126/science.177.4047.393

[116] F. Mosteller and J. W. Tukey. Data Analysis, including Statistics. In *The Collected Works of John W. Tukey*, volume IV: Philosophy and Principles of Data Analysis: 1965-1986, chapter 15, pages 601–720. Wadsworth & Brooks Cole, Monterey, CA, 1986. ISBN 0-534-05101-4

And statistics is an art, similarly to carpentry or cobblery: a practice involving specialized skills and know-how that are developed in apprenticeship with master artisans. Generally not ends in themselves, the statistical arts serve to extract information from data in situations of uncertainty, to enable actions and decisions in all fields of the human endeavor.

Counts

Under *Counting* (Page 38), we discussed evaluations of uncertainty for counted quantities: numbers of atoms of a particular isotope of radon, numbers of white blood cells (leukocytes) of different types, numbers of *Tyrannosaurus rex*, and numbers of tramcars.

For white blood cells, we considered a sample of 100 leukocytes comprising 4 eosinophils. If this count should be modeled as an outcome of a binomial random variable that counts the number of "successes" in 100 independent trials with probability of "success" 4/100, then the corresponding standard uncertainty will be

$$\sqrt{100 \times (4/100) \times (96/100)} = 1.96.$$

The Poisson model that approximates this binomial distribution has mean $100 \times (4/100) = 4$, hence standard deviation $\sqrt{4} = 2$.

A method proposed by Wilson [1927][117] to build confidence intervals for binomial proportions performs quite well in general.[118] For the true proportion of eosinophils, based on the aforementioned observed count of 4 in a sample of 100, it produces a 95 % confidence interval ranging from 0.013 to 0.11 (thus asymmetrical relative to the observed proportion, 0.04), obtained by executing the R command

[117] E. B. Wilson. Probable inference, the law of succession, and statistical inference. *Journal of the American Statistical Association*, 22:209–212, 1927. doi:10.2307/2276774

[118] R. G. Newcombe. Two-sided confidence intervals for the single proportion: comparison of seven methods. *Statistics in Medicine*, 17(8):857–872, 1998. doi:10.1002/(sici)1097-0258(19980430)17:8<857::aid-sim777>3.0.co;2-e

```
prop.test(x=4, n=100)$conf.int
```

The uncertainty analysis reported earlier for eosinophils (Page 39) takes two sources of uncertainty into account: sampling variability and between-examiner variability.

Sampling variability is modeled using a multinomial model (Page 176), to take into account the fact that the counts of the different types of leukocytes are like outcomes of dependent, binomial random variables.

Between-examiner variability is modeled using Gaussian distributions (one for each kind of leukocyte), all with mean zero and with standard deviations that depend on the type of leukocyte, and are set equal to the standard uncertainties that Fuentes-Arderiu et al. [2007] evaluated. These Gaussian "errors" are added to the counts simulated using the multinomial distribution, using a Monte Carlo method.

Bootstrap

The statistical bootstrap is a computationally-intensive method for statistical inference, and in particular for uncertainty evaluation.[119] Diaconis and Efron [1983] provide a compelling, accessible introduction to the bootstrap, and Hesterberg [2015] describes bootstrapping techniques, copiously illustrated with examples.

[119] B. Efron and R. J. Tibshirani. *An Introduction to the Bootstrap*. Chapman & Hall, London, UK, 1993; and A. C. Davison and D. Hinkley. *Bootstrap Methods and their Applications*. Cambridge University Press, Cambridge, UK, 1997. ISBN 0-521-57471-4. URL `statwww.epfl.ch/davison/BMA/`

There are two main versions of the bootstrap: parametric and non-parametric. Both can be applied to univariate and multivariate data (for example, for the scores in the Pairs Figure Skating competition of the 2022 Winter Olympics (Page 68), and for the calibration of a GC-MS instrument (Page 98) used to measure concentration of chloromethane). Here we begin with a set of replicated determinations $x_1, \dots, x_m$ of a scalar quantity, obtained under conditions of repeatability.

THE PARAMETRIC BOOTSTRAP regards these determinations as if they were a sample from a probability distribution P_θ that is indexed by a possibly multidimensional parameter θ. The underlying assumption is that this distribution is an adequate model for the variability of the replicates. We also assume that the true value of the measurand, $\eta = \psi(\theta)$, is a known function ψ of θ. The parametric bootstrap involves three steps:

(PB1) Estimate θ from $\{x_i\}$, obtaining $\widehat{\theta}$. Here we are pretending that $\widehat{\theta}$ is θ since θ itself is unknown. Most commonly, $\widehat{\theta}$ is the maximum likelihood estimate.

(PB2) Draw a large number, K, of samples of size m from $P_{\widehat{\theta}}$, and compute the estimate of θ for each of these samples, obtaining $\theta_1^*, \dots, \theta_K^*$. K should be no smaller than 10^3 when the method is used to compute standard deviations of functions of the data, and ideally of the order of 10^6 for most purposes.

(PB3) Compute the corresponding estimates of the measurand, $y_1 = \psi(\theta_1^*), \dots, y_K = \psi(\theta_K^*)$, and use them as if they were a sample drawn from the distribution of the measurand, to evaluate the associated uncertainty.

The standard deviation of the $\{y_k\}$ is an evaluation of standard uncertainty of η, and the 2.5th and 97.5th percentiles of the $\{y_k\}$ are the endpoints of a coverage interval for the true value of the measurand (η), with 95 % probability.

The parametric bootstrap is used below (Page 195) to evaluate the uncertainty associated with the maximum likelihood estimate of the tensile strength of alumina coupons in a 3-point flexure test.

THE NON-PARAMETRIC BOOTSTRAP requires that some recipe (R) be available to combine the replicated observations and to produce an estimate of the measurand: $y = R(x_1, \ldots, x_m)$. This recipe may be as simple as computing their median, or it may be an arbitrarily complicated, nonlinear function of the data. The observations again are regarded as a sample from some probability distribution, but here this distribution remains unspecified (hence the qualifier *non-parametric*).

The non-parametric bootstrap is even bolder than the parametric one. For the parametric bootstrap we estimated a parameter of a probability distribution, and proceeded to sample from this distribution pretending that the estimate of the parameter is equal to the true value of the parameter. For the non-parametric bootstrap we will treat the set of replicates in hand as if it were an infinitely large sample from the unspecified, underlying probability distribution, by taking these steps:

Step (NPB1) means that we get s_{1k} by drawing one of the observations we have as if drawing a ball from a lottery bowl, and then return it back to the bowl, mix the contents, and then draw the observation that will become s_{2k} and so on. Note that the same observation may appear multiple times in a bootstrap sample.

(NPB1) Select a large, positive integer K, and for each $k = 1, \ldots, K$ draw $s_{1k}, \ldots, s_{mk}$ uniformly at random, and with replacement, from the set $\{x_1, \ldots, x_m\}$. Each s_{ik} is equal to one of the $\{x_i\}$. For each k, the $\{s_{ik}\}$ are called a *bootstrap sample*.

(NPB2) For each bootstrap sample, compute the corresponding estimate of the measurand, $y_k = R(s_{1k}, \ldots, s_{mk})$, and then use the $\{y_k\}$ to evaluate the associated uncertainty, similarly to how it was done in step (PB3).

The number K should be as large as practicable, the guidelines being the same as offered above, for the parametric bootstrap. When applying the bootstrap, the first thing to do is to examine the probability distribution of the bootstrap estimates of the measurand, $\{y_k\}$, for example by building a histogram of these values (if the measurand indeed is a scalar quantity).

If this distribution is very "lumpy", with only a few different values, then the bootstrap may not produce a reliable uncertainty evaluation. This may happen when the number m of observations is small, or when the way of combining them tends intrinsically to produce a small number of different values (this can happen, for example, if $R(x_1, \ldots, x_m)$ is the median of the $\{x_i\}$).

In general, m should be large enough for there to be a very large number of possible, different bootstrap samples, even if not all will produce different estimates of the quantity of interest. This can be the case even when m is surprisingly small, because given a set of m observations whose values are all different from one another, it is possible to form $\binom{2m-1}{m-1} \approx 2^{2m-1}/\sqrt{m\pi}$ different bootstrap samples using the non-parametric bootstrap.

For $m = 14$ (the number of replicated determination of the mass fraction of magnesium discussed below), the number of different bootstrap samples is already over 20 million (of course, not all of these bootstrap samples produce different estimates of the measurand). It is very unlikely that, when $m < 12$, the non-parametric bootstrap will produce reliable results even when the estimate of the measurand is highly sensitive to each single observation.

Chernick [2008] suggests that the number of observations should be at least 50. However, if the number of bootstrap samples, K, is a very small fraction of $\binom{2m-1}{m-1}$, then this may suffice for the nonparametric bootstrap to produce reliable results.

Under *Combining Replicated Observations*, we apply the non-parametric bootstrap to evaluate the uncertainty associated with the median of the Walsh averages (Hodges-Lehmann estimator), using facilities available in R package `boot` [Canty and Ripley, 2020]. Next, we illustrate the non-parametric bootstrap without resorting to these facilities, to make transparently clear what is involved.

RIVER FLOOD STAGE (S) is the height of the water surface above a reference level, and *discharge* (Q) is the volumetric flow rate. The record of yearly peak discharges in the Red River of the North, for the period 1989–2018, and the corresponding flood stages measured at Fargo, North Dakota, can be used to calibrate a relationship between flood stage and discharge, so that flood stage, which is easier to measure accurately than discharge, can be used to estimate discharge.

These data are available from the USGS at `waterdata.usgs.gov/monitoring-location/05054000`

```
# Yearly peak discharge of the Red River at Fargo
# measured at the USGS station 05054000

# Flood stage (m)
S = c(10.8,  4.7,  5.2,  5.2, 8.6, 8.1,  8.6, 8.8, 12.1,  7.6,
       6.3,  6.8, 11.2,  6.9, 6.2, 8.6, 11.3, 9.4,  5.9, 12.4,
      11.3, 11.8,  5.4, 10.1, 8.5, 5.9,  5.2, 6.8,  5.7)

# Discharge (m^3/s)
Q = c(535.2, 34.5, 74.5, 73.3, 286, 317.1, 311.5, 281.5,
      792.9, 243.8, 138.8, 159.4, 574.8, 190, 153.8, 277.8,
      563.5, 382.3, 137.1, 835.3, 600.3, 770.2, 116.7, 458.7,
      294.5, 139.3, 95.4, 160.3, 130)
z = data.frame(S=S, Q=Q)
```

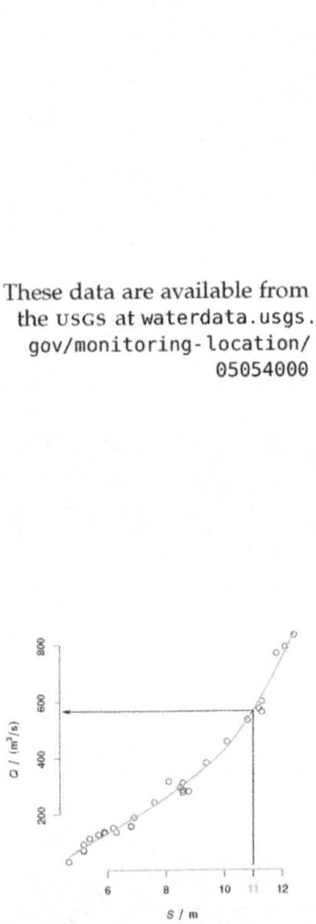

Relation between discharge (Q) and flood stage (S) for the Red River of the North, at the yearly peak discharge, for the period 1989-2018.

The following R code fits a non-parametric and locally quadratic regression model, *loess*,[120] which expresses discharge as a function of flood stage, and then uses the fitted model to estimate the discharge that corresponds to flood stage $S = 11\,\text{m}$: $\widehat{Q}(11\,\text{m}) = 567.4\,\text{m}^3/\text{s}$. The R function `predict` evaluates the associated standard uncertainty as $u(\widehat{Q}(11\,\text{m})) = 9.3\,\text{m}^3/\text{s}$.

[120] W. S. Cleveland, E. Grosse, and W. M. Shyu. Local regression models. In J. M. Chambers and T. J. Hastie, editors, *Statistical Models in S*, chapter 8. Wadsworth & Brooks/Cole, Pacific Grove, California, 1992

```
z.loess = loess(Q~S, data=z)
Q11.loess = predict(z.loess, newdata=data.frame(S=11), se=TRUE)
```

The non-parametric bootstrap, implemented below, involved drawing 10 000 samples, each of size 29, from the set of 29 pairs of observations $\{(S_i, Q_i)\}$, with replacement, fitting the *loess* model to each such sample, and then using the fitted model to predict the discharge corresponding to $S = 11\,\text{m}$. The standard deviation of the resulting 10 000 predicted values of the discharge, $13.1\,\text{m}^3/\text{s}$, is 41 % larger and a more realistic evaluation of $u(\widehat{Q}(11\,\text{m}))$ than the evaluation derived from the original *loess* fit.

```
Q11.boot = numeric(10000)
for (k in 1:10000) {
    iB = sample(1:29, size=29, replace=TRUE)
    zB.loess = loess(Q~S, data=z, subset=iB)
    Q11.boot[k] = predict(zB.loess, newdata=data.frame(S=11))
}
c(mean(Q11.boot, na.rm=TRUE),
    sd(Q11.boot, na.rm=TRUE) )
```

Combining Replicated Observations

Consider the problem of estimating the mass fraction of magnesium in a breakfast cereal, based on 14 determinations made using inductively coupled plasma optical emission spectroscopy (ICP-OES), under conditions of repeatability, which are expressed in mg/kg —

1130.0	1083.3	1091.7	1072.0	1083.2
1014.6	1068.0	1125.6	1124.6	1115.3
1088.1	1075.0	1126.8	1121.1	

These, together with other measurement results, were used to produce the certified value of the mass fraction of magnesium in NIST SRM 3233.

Choosing to minimize the mean squared difference between the estimate and the true value, or to minimize

"The problem of summarizing the location of a single batch of numbers is surely the simplest and most classical of the problems recognized as analysis of data. It was first attacked about 1580, by the use of the arithmetic mean. The next few centuries included the statement and proof of the Gauss-Markoff theorem which asserted the minimum-variance property — *among all unbiased estimates linear in the data* — in any problem where the parameters entered linearly into the average value of each observation, for the results of linear least squares. Since the use of an arithmetic mean to summarize a batch was a special instance of this general theorem, the naive might conclude that the problem of summarizing a batch had been settled. Far from it."
— John W. Tukey (1986)

the absolute value of this difference, are different options that can be interpreted as means to achieve optimal estimation under different assumptions: that these determinations are either a sample from a Gaussian distribution, or a sample from a Laplace distribution. The former suggests the arithmetic mean, the latter the median. However, many other modeling choices are conceivable, each leading to a different estimate.

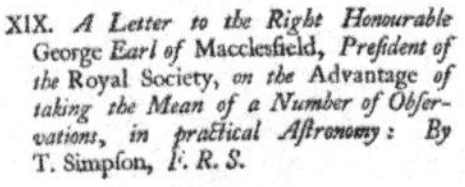
XIX. *A Letter to the Right Honourable* George *Earl of* Macclesfield, *Prefident of the* Royal Society, *on the* Advantage *of taking the Mean of a Number of Obfervations, in practical Aftronomy:* By T. Simpfon, *F. R. S.*

Thomas Simpson, Professor of Mathematics at the Royal Academy at Woolwich, outlined the advantages of averaging observations. For example, the probability that the average of six observations will have a larger absolute error than a single observation is only 25 % when the errors follow Gaussian distribution. [Simpson, 1755] (CREDIT: archive.org).

THE SIMPLE AVERAGE, or arithmetic mean, is the optimal estimate if one chooses to gauge performance in terms of mean squared error (Page 167), and if one judges the following model to be adequate for the observations: $w_i = \omega + \varepsilon_i$ for $i = 1, \ldots, m$, where $m = 14$ is number of observations, ω is the true value of that mass fraction, and the $\{\varepsilon_i\}$ are measurement errors regarded as a sample from a Gaussian distribution with mean 0 and standard deviation σ.

The statistical model, as just formulated, involves the assumption that the observations are not persistently offset from the true value they aim to estimate. This is formalized in their mathematical expectation being equal to the true value:

$$\mathbb{E}(W_i) = \mathbb{E}(\omega) + \mathbb{E}(\varepsilon_i) = \omega,$$

because ω is a constant, and the assumption was made above that $\mathbb{E}(\varepsilon_i) = 0\,\text{mg/kg}$. Note that here we have used W_i, the uppercase version of w_i, to denote the random variable that the observation w_i is regarded as a realized value of. Since the expected value of each W_i is ω, we say that there is no *bias* (persistent, or systematic error, Page 167) in the measurement.

The assumption that the measurement errors $\{\varepsilon_i\}$ are Gaussian implies that so are the $\{w_i\}$, which can be tested. The Shapiro-Wilk [Shapiro and Wilk, 1965] and the Anderson-Darling [Anderson and Darling, 1952] tests, for conformity of a sample with a Gaussian distribution, are commonly used: in this case, the former yields a p-value of 0.06, and the latter of 0.1.

It is a common convention in science that only p-values smaller than 0.05 indicate a statistically significant discrepancy, but this is a matter of (subjective) judgment (Page 32). Indeed, one cannot identify a single universal threshold of statistical significance, and some argue that the level of significance should be set at 0.005.[121]

[121] V. E. Johnson. Revised standards for statistical evidence. *Proceedings of the National Academy of Sciences*, 110:19313–19317, 2013. doi:10.1073/pnas.1313476110

Suppose that, to test a hypothesis H (in a significance test) one rejects H when the value of some test criterion (a suitable function of the data) is too large. The p-value of the test is the probability, computed on the assumption that H is true, of observing a value of the test criterion at least as large as the value that was obtained using the data available for the test (Page 32). Since a small p-value suggests that the data are unlikely if H is true, the common practice is to reject H in such case. Of course, one needs to decide in advance how small the p-value needs to be to warrant rejecting H.

The median of the observations is responsive to choices different from those that suggest the average. That instead of seeking to minimize mean squared error (Page 167), one wishes to minimize mean absolute error, which may be particularly appropriate when the measurement errors $\{\varepsilon_i\}$ have a probability distribution with heavier tails than the Gaussian: for example, Laplace (also known as double exponential, Page 172). The median of a set of observations is found by ordering the observations from smallest to largest, and selecting the middlemost (when the number of observations is odd), or the average of the two middlemost ones (when the number of observations is even).

The average of the determinations listed above, of the mass fraction of magnesium in a breakfast cereal, is 1094.2 mg/kg, and the median is 1089.9 mg/kg. The average has one serious shortcoming: it offers no protection against the influence of a single value that, for one reason or another, lies far from the bulk of the others. Suppose that, owing to a clerical error, the last value is reported as 11 211 mg/kg instead of 1121.1 mg/kg. In consequence, the average will shoot up to 1814.9 mg/kg, while the median stays put at 1089.9 mg/kg.

But the median is also open to criticism. First, it seems to gloss over most of the information in the data: it uses the data only to the extent needed to determine which is the middlemost value. Second, it is sensitive to small perturbations of the middlemost observations. Suppose that the last two digits of the third determination, 1091.7 mg/kg, are transposed accidentally, and 1097.1 mg/kg is reported instead. The average hardly budges, becoming 1094.6 mg/kg, while the median slides to 1092.6 mg/kg.

[122] J. L. Hodges and E. L. Lehmann. Estimates of location based on rank tests. *The Annals of Mathematical Statistics*, 34 (2):598–611, June 1963. doi:10.1214/aoms/1177704172

THE MEDIAN OF THE WALSH AVERAGES (better known as the Hodges-Lehmann estimate[122]) affords a fairly general, flexible solution to the problem of combining replicated observations. It is computed by taking these three steps for a sample of size m:

The Walsh averages are these: $\{(w_i + w_j)/2 : 1 \leqslant i \leqslant j \leqslant m\}$, thus including averages like $(1130.0 + 1130.0)/2$ and $(1130.0 + 1083.3)/2$, but not both $(1083.3 + 1130.0)/2$ and $(1130.0 + 1083.3)/2$, because $\{1130.0, 1083.3\}$ and $\{1083.3, 1130.0\}$ are the same subset.

(1) Compute the averages of all different subsets with two observations each (since two subsets are identical if they have the same elements regardless to order, there are $\frac{1}{2}m(m-1)$ such subsets);

(2) Form a set with these averages together with the m observations;

(3) Find the median of the $\frac{1}{2}m(m+1)$ values in this set.

The Hodges-Lehmann estimate is particularly attractive, and an excellent, general purpose replacement for the average and the median, particularly when the replicated observations may be assumed to be a sample from a symmetrical distribution, because:

- It uses the information in the data almost efficiently as the average, when the average is at its best;
- It can use the information in the data far more efficiently than the average when the average is not at its best;
- It is resistant to outliers; and
- Its standard uncertainty, as well as expanded uncertainties and coverage intervals for different coverage probabilities, can be computed very easily, for example using R.

```
w = c(1130.0, 1083.3, 1091.7, 1072.0, 1083.2, 1014.6,
      1068.0, 1125.6, 1124.6, 1115.3, 1088.1, 1075.0,
      1126.8, 1121.1)

w68 = wilcox.test(w, conf.int=TRUE, conf.level=0.68)
HL = w68$estimate; names(HL) = NULL
uHL = diff(w68$conf.int)/2

w95 = wilcox.test(w, conf.int=TRUE, conf.level=0.95)
U95HL = diff(w95$conf.int)/2
Lwr95 = w95$conf.int[1]; Upr95 = w95$conf.int[2]

c("HL"=HL, "u(HL)"=uHL, "U95(HL)"=U95HL,
  "Lwr95"=Lwr95, "Upr95"=Upr95)
```

For the 14 replicates of the mass fraction of magnesium, the median of the Walsh averages is 1098.8 mg/kg, with standard uncertainty 9.4 mg/kg, and expanded uncertainty for 95 % coverage of 18 mg/kg. Their counterparts for the average are 1094.2 mg/kg, 8.6 mg/kg, and 19 mg/kg, respectively.

And for the median, using the non-parametric statistical bootstrap (Page 180) as implemented in the following R code, we get standard uncertainty 14 mg/kg and expanded uncertainty 24 mg/kg:

```
miB = replicate(1e5, median(sample(w, 14, replace = TRUE)))
U95 = diff(quantile(miB, c(0.025,0.975)))/2
c("u(median)"=sd(miB), "U95(median)"=U95)
```

Weighted averages may be appropriate under the same general conditions that make the average optimal, but when the different observations being combined have different uncertainties, for example in the case of the determinations of equivalent activity reported for ^{59}Fe in a key comparison organized by the BIPM.[123] The synthetic radionuclide ^{59}Fe has half-life of 44.5 days, and decays to stable ^{59}Co via beta decay.

[123] C. Michotte, G. Ratel, S. Courte, K. Kossert, O. Nähle, R. Dersch, T. Branger, C. Bobin, A. Yunoki, and Y. Sato. BIPM comparison BIPM.RI(II)-K1.Fe-59 of activity measurements of the radionuclide ^{59}Fe for the PTB (Germany), LNE-LNHB (France) and the NMIJ (Japan), and the linked APMP.RI(II)-K2.Fe-59 comparison. *Metrologia*, 57(1A):06003, January 2020. doi:10.1088/0026-1394/57/1a/06003

LAB	YEAR	ACTIVITY /kBq
IAEA/RCC	1978	14 663(24)
NPL	1979	14 668(55)
ANSTO	1980	14 548(54)
CMI-IIR	1984	14 709(36)
BARC	1998	14 511(28)
KRISS	1999	14 728(50)
BKFH	2001	14 685(32)
NIST	2001	14 641(60)
PTB	2012	14 609(25)
LNE-LNHB	2013	14 603(36)
NMIJ	2014	14 576(23)

Selected measurement results for equivalent activity, A_e, of ^{59}Fe from a continuous long-term interlaboratory study [Michotte et al., 2020].

The weighted average of values $x_1, \dots, x_m$, with non-negative weights $w_1, \dots, w_m$ (which do not have to sum to 1) is

$$\overline{x}_w = \frac{x_1 w_1 + \cdots + x_m w_m}{w_1 + \cdots + w_m}.$$

If the $\{x_i\}$ are modeled as outcomes of uncorrelated random variables with a common mean μ and standard uncertainties $\{u(x_i)\}$, then the weighted average corresponding to the weights $w_i = 1/u^2(x_i)$ has smallest standard uncertainty given by

$$u_C(\overline{x}_w) = \frac{1}{\sqrt{\dfrac{1}{u^2(x_1)} + \cdots + \dfrac{1}{u^2(x_m)}}},$$

where the subscript "C" emphasizes that $u_C(\overline{x}_w)$ involves the assumption of a common mean and does not take into account how dispersed the $\{x_i\}$ actually are around their weighted average $\overline{x}_w$.

Raymond Birge was keenly aware of the fact that, in many practical situations, independent estimates of the same quantity can be markedly more dispersed than their associated standard uncertainties suggest that they should be.

For example, the standard deviation of the selected measured values listed above, of the equivalent activity of ^{59}Fe, is 68 kBq, while the median of their associated standard uncertainties is 36 kBq: if those 11 values indeed measure the same true equivalent activity unbiasedly, then that standard deviation and this median should agree *except for statistical fluctuations.*[124]

[124] R. T. Birge. The calculation of errors by the method of least squares. *Physical Review*, 40:207–227, April 1932. doi:10.1103/PhysRev.40.207

The evaluation of $u(\overline{x}_w)$ shown below is based on the weighted differences between the $\{x_i\}$ and their weighted average $\overline{x}_w$:

$$u_{\mathrm{I}}(\overline{x}_w) = \sqrt{\frac{w_1(x_1 - \overline{x}_w)^2 + \cdots + w_m(x_m - \overline{x}_w)^2}{m(w_1 + \cdots + w_m)}},$$

where m denotes the number of observations that $\overline{x}_w$ is based on. The subscript "I" refers to the fact that $u_{\mathrm{I}}(\overline{x}_w)$ entertains the possibility that the measurement results (measured values and associated uncertainties) may be mutually inconsistent: that is, that the $\{x_i\}$ may be overdispersed by comparison with what the $\{u(x_i)\}$ suggest that they should be.

The weighted average of the measured values of the equivalent activity of ^{59}Fe is 14 619 kBq. The evaluation of its standard uncertainty on the assumption that the measured values have a common mean is 10 kBq, while the evaluation based on the weighted standard deviation is 19 kBq.

The marked difference between $u_{\mathrm{C}}(\overline{x}_w)$ and $u_{\mathrm{I}}(\overline{x}_w)$ is attributable to the measurement results being mutually inconsistent, exhibiting substantial dark uncertainty (explained under *Consensus Building*, Page 146).

In such case, provided the $\{x_i\}$ can reasonably be regarded as outcomes of Gaussian random variables (possibly with different means and different standard deviations), then a classical (non-Bayesian) manner of combining them involves the application of restricted maximum likelihood estimation (REML) to fit a model that accommodates the possibility of the measured values being variously biased: $x_i = \mu + \lambda_i + \varepsilon_i$, as discussed under *Consensus Building* (Page 146).

The REML estimate for the activity of ^{59}Fe is 14 628 kBq, with associated standard uncertainty 21 kBq. It was obtained as follows:

```
Fe59.A  = c(14663, 14668, 14548, 14709, 14511, 14728,
            14685, 14641, 14609, 14603, 14576)
Fe59.uA = c(24, 55, 54, 36, 28, 50, 32, 60, 25, 36, 23)
require(metafor)
Fe59.reml = rma(yi=Fe59.A, sei=Fe59.uA, method="REML")
c("A"=Fe59.reml$b, "u(A)"=Fe59.reml$se)
```

WEIGHTED MEDIANS are preferable to the simple median when the observations being combined have different uncertainties, and the median is appropriate to begin with. The function `weighted.median` defined in package `spatstat`[125] offers a reliable implementation of the weighted median. It yields 14 606 kBq as estimate of the equivalent activity of ^{59}Fe. The associated standard uncertainty, computed using the parametric statistical bootstrap, is 17 kBq.

[125] A. Baddeley and R. Turner. spatstat: An R package for analyzing spatial point patterns. *Journal of Statistical Software*, 12:1–42, 2005. URL `www.jstatsoft.org/v12/i06/`

The following table summarizes the several estimates of equivalent activity of ^{59}Fe presented above, and their associated standard uncertainties. It should be noted that the number of measured values, $n = 11$, is too small confidently to apply the nonparametric bootstrap: therefore, only the results from the parametric bootstrap are listed, which will involve sampling from Laplace distributions: this yields the estimate 14 606 kBq, with associated standard uncertainty 17 kBq (based on 10 000 bootstrap samples).

METHOD	$A_e(^{59}\text{Fe})$	$u(A_e(^{59}\text{Fe}))$ /kBq
Average	14 631	21
Weighted Average (u_C)	14 619	10
Weighted Average (u_I)	14 619	19
Weighted Median (Par. Boot.)	14 606	17
Consensus (Gaussian REML)	14 628	21

Maximum Likelihood Estimation

Maximum Likelihood Estimation (MLE) is a technique used to estimate the values of parameters that appear in statistical measurement models. MLE may be used to estimate an input quantity that appears in a conventional measurement model as specified in the GUM, based on replicated observations, or it may be used to estimate the output quantity if the measurement model lends itself to such treatment.

We will use the initialism MLE to denote "maximum likelihood estimation," "maximum likelihood estimator," or "maximum likelihood estimate," depending on the context.

MLE produces not only an estimate of the quantity of interest, but can, in most cases, also produce an approximate evaluation of the associated uncertainty. And if supplemented with the statistical bootstrap, when this is practicable, then it can characterize uncertainty much more accurately than the approximation to standard uncertainty described in the GUM. The idea of supplementing MLE with the bootstrap is illustrated below, in relation with the measurement of the tensile strength of alumina.

MLE can be used whenever there is an explicit relationship between the true value of the quantity one wishes to estimate, and the parameters of the probability distribution of the data that is used for the purpose.

For example, when the replicated observations are from a Gaussian distribution, and the true value of the quantity of interest is the mean of this distribution. Likewise, in the example presented below, the quantity of interest (the mean tensile strength of alumina) is an explicit function of the two parameters of the Weibull distribution used to model replicated observations of the stress at which coupons of alumina break in a flexure test.

In its most succinct and general form, a statistical measurement model comprises these two statements:

$$(1) \quad X \sim P_\theta,$$

$$(2) \quad \eta = \varphi(\theta),$$

where $X = (X_1, \dots, X_n)$ is a vector of random variables whose probability distributions characterize their uncertainties. Statement (1) says that the joint probability distribution of these random variables is P_θ, where the true value of the parameter θ (typically also a vector, but with a number of components that does not vary with n) is an unknown element of a set H. Statement (2) says that η, denoting the true value of the measurand (which may be a vector), is a known function φ of θ.

Now, suppose that P_θ has probability density (Page 159) p_θ, and that x is the observed value of the vector X. The MLE of θ is $\widehat{\theta}$ that maximizes $p_\theta(x)$ as θ ranges over H:

the idea is to choose a value for the parameter θ that makes the data "most likely." The MLE of the measurand is $\widehat{\eta} = \varphi(\widehat{\theta})$.

In this process, x is kept fixed at its observed value, while θ is allowed to vary over H until a maximum of $p_\theta(x)$ is found. To emphasize this fact, one often defines a function L_x, called the *likelihood function*, as follows: $L_x(\theta) = p_\theta(x)$. None of the pieces changes, only the viewpoint: the subscript x in L_x is a way of saying that L_x depends on x but that x remains fixed while we seek to maximize $L_x(\theta)$ by varying its argument, θ, over the set H of its possible values. In applications, the subscript x is often suppressed because the dependence on x is understood, and one writes simply $L(\theta)$.

Therefore, maximum likelihood estimation amounts to maximizing the likelihood function. In some cases this can be done analytically, based on the first and second derivatives of $\ln L_x$ with respect to θ. In other cases it has to be done via numerical optimization.

Under very general circumstances, maximum likelihood estimation enjoys several remarkable properties that, coupled with the ease with which the MLE can be computed, make this method of estimation a very attractive, general purpose technique. These properties include:

- MLE produces the estimate, $\widehat{\theta}$, of the measurand with smallest uncertainty;
- The probability distribution of $\widehat{\theta}$ (which is the value of a random variable because it is a function of the data) is approximately Gaussian, and the quality of the approximation improves as the number, n, of inputs increases;
- The inverse of the matrix of second-order partial derivatives of $-\ln L_x$ with respect to θ, evaluated at $\theta = \widehat{\theta}$, is an approximation to the covariance matrix of $\widehat{\theta}$. The larger the sample that $\widehat{\theta}$ is based on, the better the approximation.

MAXIMUM LIKELIHOOD ESTIMATION OF THE WEIBULL DISTRIBUTION is applied here to characterize the tensile strength of alumina coupons based on 30 observations, made under conditions of repeatability, of the rupture stress of the coupons in a 3-point flexure test.

RUPTURE STRESS, σ/MPa					
307	407	435	455	486	371
409	437	462	499	380	411
441	465	499	393	428	445
466	500	393	430	445	480
543	402	434	449	485	562

Rupture stress for 30 alumina coupons in a 3-point flexure test. Courtesy of George D. Quinn (Material Measurement Laboratory, NIST).

The model selected for the variability of these determinations is the Weibull probability distribution (Page 171), which has two parameters that, in the present context, are called the *characteristic strength* σ_C, and the *Weibull modulus*, m.[126] Note that, throughout this example, the Greek letter σ is used to denote stress (with the same units as pressure), not standard deviation.

[126] J. B. Quinn and G. D. Quinn. A practical and systematic review of Weibull statistics for reporting strengths of dental materials. *Dental Materials*, 26:135–147, 2010. doi:10.1016/j.dental.2009.09.006

Three-point flexural strength test of an alumina coupon, light colored, between the rollers held by rubber bands.

Consistently with the notation used for the general description of the MLE above, we should then write $\theta = (m, \sigma_C)$. The measurand is the tensile strength $\eta = \sigma_C \Gamma(1 + 1/m)$, which is the mean of that Weibull distribution (and Γ is the gamma function).

The Weibull probability distribution has the following probability density: (Page 159):

$$p(\sigma_i \mid m, \sigma_C) = \frac{m}{\sigma_C} \left(\frac{\sigma_i}{\sigma_C} \right)^{m-1} e^{(-\sigma_i/\sigma_C)^m},$$

where the scale parameter σ_C and the shape parameter m are positive quantities.

Assuming that the $n = 30$ replicates of σ are like outcomes of independent Weibull random variables, the likelihood function is L such that

$$L_\sigma(m, \sigma_C) = \prod_{i=1}^{n} p(\sigma_i \mid m, \sigma_C).$$

The maximum likelihood estimates of the parameters are the values of m and σ_C that maximize $L_\sigma(m, \sigma_C)$ as a function of m and σ_C, with $\sigma = (\sigma_1, \ldots, \sigma_n)$ kept fixed at the observed rupture stresses.

Since $L_\sigma(m, \sigma_C)$ is a product of terms involving m and σ_C, it is generally preferable to maximize $\ln L_\sigma(m, \sigma_C)$ instead. The reason is that the gradient of a sum is generally better behaved during numerical optimization than the gradient of a product because the second derivatives of a sum generally do not change too much or too rapidly.

The R function `optim` minimizes the value of the function `negLogLik` with respect to its argument, the vector `par`, whose elements are the Weibull parameters, using the Nelder-Mead method [Nelder and Mead, 1965]. It requires that initial guesses be provided for the values of the parameters. The code requests that the matrix of second-order partial derivatives (Hessian matrix, named after Ludwig Otto Hesse, 1811–1874) be computed and returned because its inverse is an approximation to the covariance matrix of the parameter estimates. The larger the sample size, which is 30 in this case, the better the approximation.

The R code below minimizes the negative log-likelihood function, $-\ln L_\sigma(m, \sigma_C)$, which is equivalent to maximizing the likelihood function.

```
sigma = c(307, 371, 380, 393, 393, 402, 407, 409, 411, 428,
          430, 434, 435, 437, 441, 445, 445, 449, 455, 462,
          465, 466, 480, 485, 486, 499, 499, 500, 543, 562)
negLogLik = function(par, s = sigma) {
  -1 * sum(dweibull(s, shape=par[1], scale=par[2], log=TRUE)) }
## Find maximum likelihood estimates
opt = optim(par = c(m=10.6, sigmaC=465), fn = negLogLik,
            s = sigma, hessian = TRUE)
## Estimates of the shape and scale parameters
opt$par
## Approximate covariance matrix of the parameter estimates
V = solve(opt$hessian)
## Approximate standard uncertainties of the parameter estimates
sqrt(diag(V))
```

The results are $\widehat{m} = 9.24$, $\widehat{\sigma}_C = 467\,\text{MPa}$, hence $\widehat{\eta} = 443\,\text{MPa}$. The last line of the previous R code will produce approximate evaluations of $u(\widehat{m}) = 1.23$ and $u(\widehat{\sigma}_C) = 9.8\,\text{MPa}$.

To compute $u(\widehat{\eta})$ one can use the fact that the equation $\eta = \sigma_C\Gamma(1 + 1/m)$ is the measurement model, while recognizing that $\widehat{m}$ and $\widehat{\sigma}_C$ are correlated. The correlation between them is 0.33, which can be obtained using `cov2cor(V)` following the R code above. The *NIST Uncertainty Machine* then yields $u(\widehat{\eta}) = 10.4\,\text{MPa}$.

These uncertainty evaluations are made possible by the aforementioned MLE magic. However, this magic requires a large number of observations, while we have only 30. May this be enough?

To answer this question without invoking the MLE magic, we can redo the uncertainty analysis employing the parametric statistical bootstrap [Efron and Tibshirani, 1993] (Page 179), and compare the evaluations we will get this way with those we got above.

The idea is to take the above MLEs of m and σ_C and use them to generate many samples of size 30 from the Weibull distribution with these values of the parameters. For each such sample, we find the best parameter values by minimizing the negative log-likelihood, $-\ln L_\sigma(m, \sigma_C)$.

```
m.HAT = opt$par['m']
sigmaC.HAT = opt$par['sigmaC']
boot = array(dim=c(1e5, 3))
colnames(boot) = c('m', 'sigmaC', 'eta')
for (j in 1:1e5) {
  sigmaB = rweibull(30, shape=m.HAT, scale=sigmaC.HAT)
  thetaB.MLE = optim(par=c(m=10, sigmaC=440),
                     fn=negLogLik, s=sigmaB)$par
  ## Calculate eta
  etaB = thetaB.MLE['sigmaC']*gamma(1 + 1/thetaB.MLE['m'])
  boot[j,] = c(thetaB.MLE, etaB)
}
apply(boot, 2, sd)
```

This R code produces $u(\widehat{m}) = 1.47$, $u(\widehat{\sigma_C}) = 9.8\,\text{MPa}$, and $u(\widehat{\eta}) = 10.5\,\text{MPa}$. Not only does this exercise validate the MLE magic in this case, it also gives us the ingredients to characterize the joint probability distribution of $\widehat{m}$ and $\widehat{\sigma}_C$, hence also the distribution of $\widehat{\eta}$.

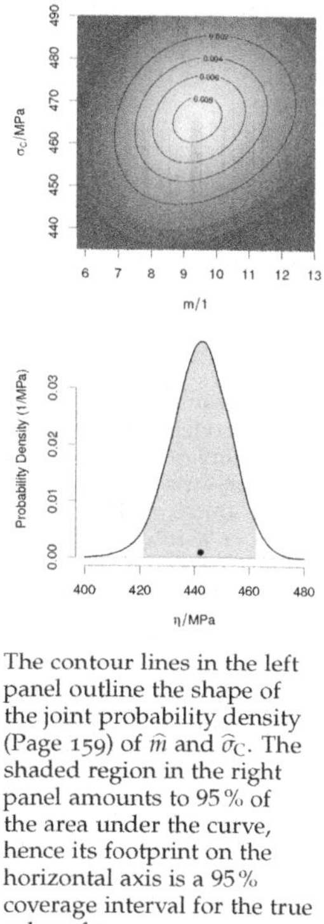

The contour lines in the left panel outline the shape of the joint probability density (Page 159) of $\widehat{m}$ and $\widehat{\sigma}_C$. The shaded region in the right panel amounts to 95 % of the area under the curve, hence its footprint on the horizontal axis is a 95 % coverage interval for the true value of η.

Least Squares

Least squares is a criterion of estimation that is often also described as a method for the adjustment of observations.

Consider the simplest instance of such adjustment, where one has made m replicated determinations of the same quantity, $x_1, \ldots, x_m$, which one wishes to combine by choosing the value θ that minimizes the sum of squared deviations of the observations from it:

$$S(\theta) = (x_1 - \theta)^2 + \ldots + (x_m - \theta)^2.$$

Such θ is the solution of $S'(\theta) = 0$, where S' denotes the first derivative of S with respect to θ, hence reduces to $(-2)(x_1 - \theta) + \ldots + (-2)(x_m - \theta) = 0$. Solving this equation for θ yields

$$\theta = (x_1 + \cdots + x_m)/m = \overline{x},$$

the average of the observations. This is indeed the value where $S(\theta)$ achieves its minimum because.

$$S''(\theta) = 2m > 0.$$

If measurement errors are best modeled using a probability distribution other than Gaussian, then an adjustment of observations based on a different criterion may be preferable. For example, minimizing the sum of the absolute values of the errors will lead to the *median*, which is the maximum likelihood solution when the errors follow a Laplace distribution (Page 172).

If the measurement errors follow the Gaussian distribution, then least squares is equivalent to maximum likelihood estimation.

The method of least squares was developed by Adrien-Marie Legendre (1752–1833) and Carl Friedrich Gauss (1777–1855) at the beginning of the 19th century. In an early, and most remarkable application of this method, Gauss predicted where the asteroid Ceres should be found again after it had last been observed by its discoverer, Giuseppe Piazzi (1746–1826).[127] And it was indeed at the location predicted by Gauss that Franz Xaver von Zach (1754–1832) and Heinrich Olbers (1758–1840) spotted Ceres in the skies on the last day of 1801.

[127] C. F. Gauss. Summarische Uberficht der zur bestimmung der bahnen der beyden neuen hauptplaneten augewanden methoden. *Monatliche Correspondenz zur Beförderung der Erd- und Himmels- Kunde*, XX(Part B, July-December, Section XVII): 197–224, September 1809

The method of least squares can be illustrated with an

example we encountered earlier (Page 61), to determine the mass of three weights by measuring their mass differences. This example involves three observations of mass differences ($D_{AB} = -0.38\,\text{mg}$, $D_{AC} = -1.59\,\text{mg}$, and $D_{BC} = -1.22\,\text{mg}$), three parameters whose values we are seeking (true masses of all weights, that is, $200\text{ g} + \delta_A$, $200\text{ g} + \delta_B$, and $200\text{ g} + \delta_C$), and a constraint $K = \delta_A + \delta_B = 0.83\,\text{mg}$ that must be satisfied while also taking into account its associated uncertainty, $u(K) = (0.07\,\text{mg}) \times \sqrt{2}$.

The three observations are mutually inconsistent because, for example, $D_{AB} - D_{AC} = -1.21\,\text{mg}$ while $D_{BC} = -1.22\,\text{mg}$. To make them consistent we introduce non-observable "errors" ε_1, ε_2, and ε_3, such that the following three equations hold true simultaneously

$$\begin{aligned} D_{AB} &= \delta_A - \delta_B + \varepsilon_1, \\ D_{AC} &= \delta_A - \delta_C + \varepsilon_2, \\ D_{BC} &= \delta_B - \delta_C + \varepsilon_3. \end{aligned}$$

Applying the method of least squares in this case amounts to choosing values for δ_A, δ_B, and δ_C that minimize the sum of the squared errors, $\varepsilon_1^2 + \varepsilon_2^2 + \varepsilon_3^2$, while also satisfying the constraint

$$K = \delta_A + \delta_B = 0.83\,\text{mg}.$$

This constraint is "soft" because it is surrounded by uncertainty, $u(K) = (0.07\,\text{mg}) \times \sqrt{2}$. However, let us first pretend that it is "hard" so that we can replace δ_B with $K - \delta_A$ and write the optimization criterion as follows:

$$\begin{aligned} S(\delta_A, \delta_C) &= \varepsilon_1^2 + \varepsilon_2^2 + \varepsilon_3^2 \\ &= (D_{AB} - \delta_A + (K - \delta_A))^2 + (D_{AC} - \delta_A + \delta_C)^2 \\ &\quad + (D_{BC} - (K - \delta_A) + \delta_C)^2. \end{aligned}$$

The values of δ_A and δ_C that minimize $S(\delta_A, \delta_C)$ correspond to a situation when both partial derivatives equal zero,

$$\partial S(\delta_A, \delta_C)/\partial \delta_A = 0 \text{ and } \partial S(\delta_A, \delta_C)/\partial \delta_C = 0,$$

These estimates, $\widehat{\delta}_A$ and $\widehat{\delta}_C$, indeed correspond to a minimum of the criterion because the matrix of second order partial derivatives of $S(\delta_A, \delta_C)$ is diagonal and both elements in its main diagonal are positive integers: $\partial^2 S/(\partial \delta_A)^2 = 12$ and $\partial^2 S/(\partial \delta_C)^2 = 4$.

that is

$$\widehat{\delta}_A = +\tfrac{1}{3}D_{AB} + \tfrac{1}{6}D_{AC} - \tfrac{1}{6}D_{BC} + \tfrac{1}{2}K = 0.227\,\text{mg},$$
$$\widehat{\delta}_C = -\tfrac{1}{2}D_{AC} - \tfrac{1}{2}D_{BC} + \tfrac{1}{2}K = 1.82\,\text{mg}.$$

Applying the constraint yields the estimate of the remaining parameter, $\widehat{\delta}_B = K - \widehat{\delta}_A = 0.603$ mg.

Now we need to bring into play the "softness" of the constraint, which is the uncertainty of K. This can be accomplished in any one of several different ways. The most intuitive one may be a Monte Carlo procedure.

The idea is to solve the same optimization problem we just solved, when we pretended that the constraint was "hard", but to do it many times over, each time using a value for the constraint drawn from a probability distribution with mean K and standard deviation $u(K)$. We will use a Gaussian distribution for this purpose, in keeping with the spirit of least squares.

```
D.AB = -0.38; D.AC = -1.59; D.BC = -1.22
abc = array(dim=c(1e6, 3))
for (i in 1:1e6) {
  k = rnorm(1, mean=0.83, sd=0.07*sqrt(2))
  A = D.AB/3 + D.AC/6 - D.BC/6 + k/2
  B = k - a
  C = -D.AC/2 - D.BC/2 + k/2
  abc[i,] = c(A, B, C) }
apply(abc, 2, mean)
apply(abc, 2, sd)
```

The final, constrained least squares estimates are

$$\widehat{\delta}_A = 0.227\,\text{mg}, \; \widehat{\delta}_B = 0.603\,\text{mg}, \text{ and } \widehat{\delta}_C = 1.82\,\text{mg},$$

with associated uncertainties

$$u(\widehat{\delta}_A) \approx u(\widehat{\delta}_B) \approx u(\widehat{\delta}_C) = 0.049\,\text{mg}.$$

More general constrained least squares problems can be solved using the method of Lagrange multipliers, as described by Zelen [1962] and Seber [2008, §24.3]. R function `solnp`, in package `Rsolnp` implements a versatile algorithm for constrained, nonlinear optimization using an augmented Lagrange method.[128]

[128] Y. Ye. *Interior Point Algorithms: Theory and Analysis*. John Wiley & Sons, New York, NY, 1997. ISBN 978-0471174202; and A. Ghalanos and S. Theussl. *Rsolnp: General Non-linear Optimization Using Augmented Lagrange Multiplier Method*, 2015. R package version 1.16

The method of least squares is very often used to fit models to data, and it is also very often misused because users fail to realize how attentive this method is toward every little detail in the data, while such solicitude may, in many cases, prove excessive. For example, a single data point that markedly deviates from the pattern defined by the others can lead the least squares fit astray.

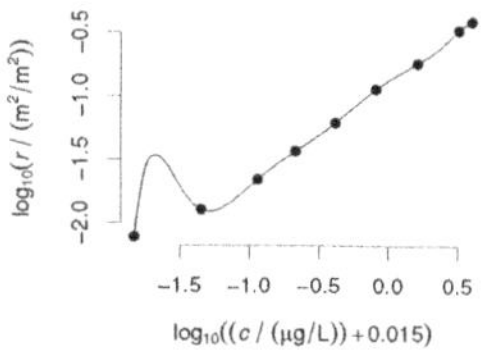

Even though a polynomial of the 8th degree fits the median values of r at each value of c (dots) exactly, it would be an unrealistic calibration function.

It may also happen that a least squares fit reproduces the data exactly yet is ridiculous. A figure presented earlier and reproduced here illustrates this point in spades. The fit, which may be computed using the R code below, goes through each data point exactly, but at the price of an odd, obviously unrealistic contortion of the curve. The residuals, which are the differences between observed and fitted values of $\log_{10}(r/(\mathrm{m}^2/\mathrm{m}^2))$, are all zero because the method of least squares forces a polynomial (regardless of degree), with as many coefficients as there are data points, to pass through all of them, at any cost.

```
x = c(-1.824, -1.347, -0.939, -0.668, -0.382,
      -0.089, 0.208, 0.507, 0.604)
y = c(-2.107, -1.892, -1.653, -1.432, -1.208,
      -0.942, -0.74, -0.476, -0.409)
summary(lm(y~poly(x, degree=8, raw=TRUE)))
```

When the method of least squares is used either to adjust observations or to fit a function to empirical data, it is often applied subject to constraints. For example, when the purpose is to adjust mass fractions of a compound whose constituents are determined separately from one another, one will wish to constrain the adjusted mass fractions to be non-negative, or to be less than 1 g/g, or to sum to 1 g/g, or possibly to satisfy more than one such constraint simultaneously. Similarly, when fitting a piecewise polynomial function to data, one may wish to constrain the result to be continuous and smooth, that is, to be a *spline* [Ferguson, 1986].

Between approximately −260 °C and 960 °C, the International Temperature Scale (ITS-90) is defined by means of platinum resistance thermometers calibrated at specified fixed points (such as melting points of various metals). In this temperature interval, the ITS-90 reference function is given by two high-order polynomials constrained to join at the triple point of water without discontinuity of either the polynomials or of their first derivatives.

Generalized Linear Models

The method of least squares (Page 196) is often used to fit *linear models* that express the mean value, η, of an observable output, y, as a linear combination of values of inputs, $x_1, \ldots, x_n$, whose values are known with negligible uncertainty:

$$\eta = \beta_0 + \beta_1 x_1 + \cdots + \beta_n x_n,$$
$$y = \eta + \varepsilon,$$

where ε denotes the measurement error affecting y.

Least squares is optimal when the measurement errors affecting the outputs that are observed at different combinations of values of the inputs, are like a sample from a Gaussian distribution with mean 0 and a finite standard deviation whose value generally is unknown. In such circumstances, the outputs have Gaussian distributions, too, but with different means.

When the outputs do not have Gaussian distributions, as was the case when we considered the probability of death for mussels (Page 125) exposed to deltamethrin, the options commonly available are either to re-express [Mosteller and Tukey, 1977, Chapters 4-6] the outputs so that a linear Gaussian model becomes tenable for them, or to employ a *generalized linear model*.[129]

[129] P. McCullagh and J. A. Nelder. *Generalized Linear Models*. Chapman & Hall / CRC, London, UK, 2nd edition, 1989

The number of dead mussels had a binomial (Page 173) probability distribution whose parameter, the probability of death, depended on the level of exposure: more precisely, the model expressed the true value of the log-odds of death as a linear function of the logarithm of the mass concentration of deltamethrin in the water where the mussels lived.

A generalized linear model (GLM) for a response y — which may be continuous or discrete — expresses η, the expected value of y, as a given function of a linear combination of the predictors,

$$\eta = \varphi^{-1}(\beta_0 + \beta_1 x_1 + \cdots + \beta_n x_n),$$

and specifies the probability distribution of y. Here, φ^{-1} denotes the mathematical inverse of the *link function* φ.

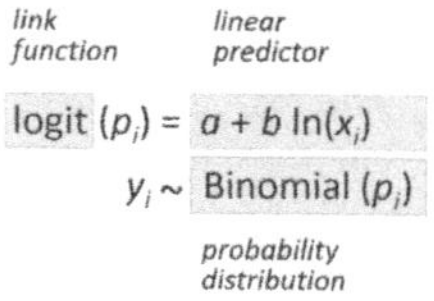

Generalized linear model.

In the case of deltamethrin example, the probability of death was $p(c)$ when the mass concentration of the toxin was c, the sole input, and

$$p(c) = \frac{\exp(\alpha + \beta \ln c)}{1 + \exp(\alpha + \beta \ln c)}.$$

Hence, the link function was the logit, which maps p to $\ln(p/(1-p))$ $(0 < p < 1)$, and y had a binomial distribution with probability of "success" $p(c)$. R function `glm` can be used to fit a wide range of generalized linear models to data by the method of maximum likelihood.

Model Selection

When we built a model for the calibration function used to measure the mass concentration of chloromethane (Page 98) we employed the Bayesian Information Criterion (BIC) as a guide to select one among several alternative models, and pointed out that the smaller the BIC, the more adequate the model. Here we describe how BIC is computed, and explain why the best model (among several under consideration) has the smallest value of BIC.

Consider fitting a straight line (1st degree polynomial)

to the chloromethane data, using the following R code:

```
x = c(-1.824, -1.347, -0.939, -0.668, -0.382,
      -0.089, 0.208, 0.507, 0.604)
y = c(-2.107, -1.892, -1.653, -1.432, -1.208,
      -0.942, -0.74, -0.476, -0.409)
summary(lm(y~poly(x, degree=1, raw=TRUE)))
```

The model treats the $m = 9$ values of x as known without uncertainty, and regards the values of y as outcomes of m independent Gaussian random variables whose means depend on the values of x. More precisely, y_i is an outcome of a Gaussian random variable with mean $\beta_0 + \beta_1 x_i$ and standard deviation σ, for $i = 1, \ldots, m$. The *likelihood function* corresponding to these data is a function L of the three parameters β_0, β_1, and σ, where the data $\{x_i, y_i\}$ are kept fixed, such that

$$L_{x,y}(\beta_0, \beta_1, \sigma) = \left(\frac{1}{\sigma\sqrt{2\pi}}\right)^m \exp\left\{-\sum_{i=1}^{m} \frac{(y_i - \beta_0 - \beta_1 x_i)^2}{2\sigma^2}\right\}.$$

In these circumstances, the maximum likelihood estimates of β_0, and β_1 are the least squares (Page 196) estimates, $\widehat{\beta}_0$ and $\widehat{\beta}_1$, and the maximum likelihood estimate of σ^2 is the average of the squared residuals $\{y_i - \widehat{y}_i\}$, where $\widehat{y}_i = \widehat{\beta}_0 + \widehat{\beta}_1 x_i$, for $i = 1, \ldots, m$: that is,

$$\widehat{\sigma}^2 = \sum_{i=1}^{m} (y_i - \widehat{y}_i)^2 / m.$$

Note that k is not the degree of the polynomial; it is the number of adjustable parameters. For polynomial regression models, like the ones we are comparing here, k is the number of coefficients of the polynomial plus the additional parameter, σ.

The BIC for this model and data is

$$\text{BIC} = -2 \ln L_{x,y}(\widehat{\beta}_0, \widehat{\beta}_1, \widehat{\sigma}) + k \ln m,$$

where $k = 3$ denotes the number of model parameters. The closer the model fits the data, the larger the value $L_{x,y}(\widehat{\beta}_0, \widehat{\beta}_1, \widehat{\sigma})$ that the likelihood function takes at the maximum likelihood estimates. Or, equivalently, the more accurate the model, the smaller (the more negative) the first term in the foregoing definition of the BIC,

$-2\ln L_{x,y}(\widehat{\beta}_0, \widehat{\beta}_1, \widehat{\sigma})$, (because it has a minus sign in front of it and the logarithm is an increasing function).

In general, the larger the number of parameters in a model, the closer it will fit the data. Therefore, the greater the degree of the polynomial, the closer the fit (above we saw that a polynomial of the 8th degree will fit these data exactly), and the smaller (the more negative) $-\ln(L)$ will be. On the other hand, since k denotes the number of parameters in the model, the larger this number the larger the second term, $k\ln(m)$, in the definition of the BIC, which is added to the first.

That is, the two terms in the BIC move in opposite directions as the number of parameters in the model increases: the first term becomes smaller, while the second increases. The first term rewards goodness of fit (the smaller the better), while the second term, $k\ln(m)$, penalizes model complexity (the larger the worse), where "complexity" here means number of adjustable parameters. In summary, when we select the model that minimizes BIC we are striking a compromise between goodness-of-fit and model complexity.

DEGREE	k	BIC
1	3	−19.9
2	4	−32.3
3	5	−43.4
4	6	−41.3
5	7	−43.1
6	8	−41.7
7	9	−41.3

The smaller the value of BIC, the more adequate the model for the data. In this case BIC decreases appreciably as the degree of the polynomial increases from 1 to 3, but then stabilizes, fluctuating around the same value. This suggests that a polynomial of the third degree may be the best choice for these data.

The following R code computes BIC for the first degree polynomial model described above. It does it both from scratch and also using the built-in function BIC.

```
x = c(-1.824, -1.347, -0.939, -0.668, -0.382,
      -0.089, 0.208, 0.507, 0.604)
y = c(-2.107, -1.892, -1.653, -1.432, -1.208,
      -0.942, -0.74, -0.476, -0.409)
y1.lm = lm(y~poly(x, degree=1, raw=TRUE))
## Size of the sample the model was fitted to
n = nrow(y1.lm$model)
## sigma is the extra parameter
k = length(y1.lm$coefficients) + 1
## MLE of sigma
sigmaHAT = sqrt(mean(residuals(y1.lm)^2))
yHAT = fitted.values(y1.lm)
loglik = sum(dnorm(y, mean=yHAT, sd=sigmaHAT, log=TRUE))
c("BIC"=-2*loglik + k*log(n), "BIC"=BIC(y1.lm))
```

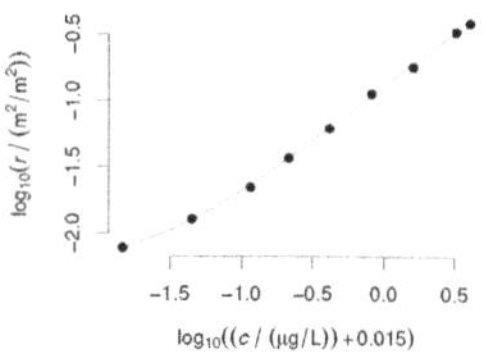

The best model according to BIC, among those under consideration, is a polynomial of the 3rd degree.

Bayesian Estimation

Bayesian estimation consists of estimating the measurand and evaluating the uncertainty associated with the estimate by application of Bayes's rule. The defining trait of Bayesian methods is to treat all unknown values of properties of interest as non-observable random variables, and the measurement results as observed values of random variables. O'Hagan [2008] provides a concise overview of Bayesian principles and methods, and Gelman et al. [2003] describe and illustrate the contemporary practice of Bayesian statistics.

The Bayesian approach provides integrated, simultaneous solutions to the problems of estimation and uncertainty evaluation, while classical solutions tend to solve these problems separately and in succession. Among classical methods, MLE comes the closest to Bayesian methods in this respect.

Bayesian methods can also be very useful in situations where there are about as many parameters as there are observations, for example in image reconstruction,[130] because they act as a regularization prescription, in the sense that the information conveyed by the prior distribution helps solve what could otherwise become an ill-posed optimization problem [Rasmussen and Williams, 2006] [Hastie et al., 2009].

The prior information may originate in similar studies carried out in the past, or it may reflect expert knowledge: in either case, it must be cast in the form of a probability distribution on the set of possible values of the measurand. When an expert is the source of prior information, one should employ a disciplined approach to elicit the relevant information and to encapsulate it in a probability distribution.[131] [132]

The Bayesian approach to draw inferences from data aligns the interpretation of such inferences with how most people are naturally inclined to interpret them.

[130] S. Geman and D. Geman. Stochastic relaxation, Gibbs distributions, and the Bayesian restoration of images. *IEEE Transactions on Pattern Analysis and Machine Intelligence*, 6:721–741, 1984

[131] A. O'Hagan, C. E. Buck, A. Daneshkhah, J. R. Eiser, P. H. Garthwaite, D. J. Jenkinson, J. E. Oakley, and T. Rakow. *Uncertain Judgements: Eliciting Experts' Probabilities*. Statistics in Practice. John Wiley & Sons, Chichester, England, 2006. ISBN 978-0-470-02999-2

[132] D. E. Morris, J. E. Oakley, and J. A. Crowe. A web-based tool for eliciting probability distributions from experts. *Environmental Modelling & Software*, 52:1–4, February 2014. doi:10.1016/j.envsoft.2013.10.010

This advantage is clearest in relation with the interpretation of coverage intervals.

The conventional interpretation, which has pervaded the teaching of statistics for at least 70 years, is as follows: a 95 % interval for the true value of a quantity is a realization of a random interval, and the 95 % probability does not apply specifically to the interval one actually gets, but is a property of the procedure that generates such interval, characterizing its performance in the long run.

This interpretation typically goes hand in hand with a frequentist interpretation of probability, which justifies statements like this: the 95 % means that, of all such intervals that a statistician produces in her lifetime, 95 % cover their intended targets, and 5 % miss them.

Bayesian statistics gets its name from the 18th century English statistician, philosopher, and minister, Thomas Bayes, whose most famous accomplishment was published only posthumously [Bayes and Price, 1763].

The Bayesian interpretation (Page 204) of coverage intervals is more intuitive, and certainly applies to the specific interval that one actually gets: the 95 % is the probability that the value of interest is in that particular interval that one has computed.

This interpretation is enabled by a change in viewpoint: the interval one gets is as concrete and definite as can be — there being nothing random about it. The "randomness" is transferred to the quantity whose true value is unknown, while the very meaning of "random" is refreshed. From a Bayesian viewpoint, a random quantity does not have a value that fluctuates unpredictably like a leaf fluttering in the wind — its value is what it is, but our knowledge of it is imperfect or incomplete.

Bayesians use probability distributions to quantify degrees of belief (in the truth of propositions about the true values of properties under study), or to describe states of partial or incomplete knowledge about properties. A random variable is simply a property (quantitative or qualitative) that has a probability distribution as an attribute. This attribute is not an intrinsic attribute of the property. Instead, it describes an epistemic relation between the person aiming to learn the true value of the property, and this true value.

The Bayesian approach is eminently practical because its specific results have the meaning common sense expects them to have, and they are immediately relevant because they are not contingent on what may happen in the rest of anyone's lifetime (refer to the discussion above of the meaning of confidence intervals).

In a nutshell, the Bayesian approach to estimation and uncertainty evaluation for statistical measurement models involves modeling all parameters whose true values are unknown as (non-observable) values of random variables, and the measurement data as observed outcomes of random variables whose distributions depend on the unknown parameter values. The estimate of the measurand, and an evaluation of the associated uncertainty, are derived from the conditional distribution of the unknowns given the data.

Combining *prior knowledge* with experimental results might seem strange at first. After all, why do we need to incorporate any prior knowledge? Why not rely entirely on the experimental data? There are several simple but compelling answers as to why, including these:

(a) More often than not there is prior knowledge about the measurand, otherwise one would not even be able to select a measuring instrument, and the Bayesian approach provides the disciplined means to take such prior knowledge into account;

(b) There may be hard or soft constraints that the parameters in the statistical model for the observations should satisfy, and probability distributions can be used to enforce the effectively, for example in the context of constrained least squares (Page 196);

(c) The ratio between the number of parameters in the model, and the number of data points, may be large enough that maximum likelihood estimation, for example, can become an ill-conditioned problem: in such cases the prior distribution acts as a regularization prescription that actually enables reliable estimation.

Consider the interpretation of test results for a disease, such as COVID-19. Suppose that a rapid test has been developed that correctly yields a positive test result in 99 % of people infected with COVID, and that yields a false positive for 1 % of the people without COVID.

If you take the test and it turns out positive, most likely you will conclude that you are infected. However, the probability of a patient having COVID given a positive test result depends not only on the the *sensitivity* (probability of correctly detecting true positives) and *specificity* (probability of identifying the target disease correctly) of the test, but also on the *prevalence* (proportion of people who are infected) of the disease!

Consider two COVID tests both done in Florida using that same method: one test is taken in late October 2020, and the other three months later. Noh and Danuser (2021) estimate that some 50 000 people were infected with COVID in late October 2020 in Florida, whereas three months later that number rose to one million.[133]

[133] J. Noh and G. Danuser. Estimation of the fraction of COVID-19 infected people in U.S. states and countries worldwide. *PLoS ONE*, 16:e0246772, 2021. doi:10.1371/journal.pone.0246772

If we incorporate this prior information about the fraction of the population infected with COVID, then the probability that the patient with the positive test result really has COVID is very different in those two periods: about 20 % in October 2020, and about 85 % in February 2021, clearly demonstrating that measurements alone might not provide all the information that should be taken into account to make the best decisions.

To show how these conclusions were reached, consider the expected counts corresponding to the foregoing probabilities. The expected number of people in Florida who were infected in October 2020 was 50 000. If all of them had been tested, then $0.99 \times 50\,000 = 49\,500$ would have tested positive.

The expected number who were not infected is approximately 20 million, based on the population of Florida. And if all of these had been tested, then $0.01 \times 20 \times 10^6 = 200\,000$ would have tested positive.

The *prior distribution* tells us the likely whereabouts of the parameters before we gather any new data. The *likelihood* tells us how likely the data are given any particular values of the parameters. Bayes's rule [DeGroot and Schervish, 2012, 2.3,7] puts these two pieces together to tell us how likely it is that the true values of the parameters will be in any specified subsets of their ranges, in light of the fresh data, and with due allowance for the prior information.

Therefore, the proportion of those that were infected among those that tested positive would have been $49\,500/(49\,500 + 200\,000) = 0.20$. Similar calculations yield the corresponding, much larger proportion, 0.85, for early February of 2021.

Bayesian estimation of the tensile strength η of alumina coupons starts from the same data and uses the same likelihood function that we used above to illustrate the method of maximum likelihood estimation.

The prior knowledge in hand consists of facts about the Weibull modulus m and the characteristic strength σ_C that have been established in previous studies of rupture of the same material, also in 3-point flexure testing: that m is around 8.8, give or take 1.25, and that σ_C is around 467 MPa give or take 11 MPa. We capture these facts by modeling m and σ_C *a priori* as independent random variables with Gaussian distributions, m with mean 8.8 and standard deviation 1.25, σ_C with mean 467 MPa and standard deviation 11 MPa. This defines the *prior distribution*, whose probability density (Page 159), π, is the product of two Gaussian probability densities, one for m, the other for σ_C.

Given any hypothetical values of m and σ_C, the observed values of rupture stress, $\sigma = (\sigma_1, \ldots, \sigma_{30})$, for the 30 coupons that were tested, are modeled as outcomes of 30 independent random variables, all with the same Weibull distribution with shape m and scale σ_C. The product of the corresponding 30 Weibull densities, each evaluated at an observed value of rupture stress, then becomes a function of m and σ_C alone (the observations of rupture stress, $\{\sigma_i\}$, are all frozen at their observed values). This is the same likelihood function, $L_\sigma(m, \sigma_C)$, that we encountered while discussing maximum likelihood estimation (Page 191).

The conditional distribution of the parameters given the data (which actually is the version of the prior distribution suitably updated by incorporation of the fresh data), the so-called *posterior distribution*, has probability density (Page 159) given by Bayes's Rule:

Referring to Bayes's Rule, Jeffreys [1973, 2.3] pointed out that "This theorem (due to Bayes) is to the theory of probability what Pythagoras's theorem is to geometry."

$$q_\sigma(m, \sigma_C) = \frac{L_\sigma(m, \sigma_C)\, \pi(m, \sigma_C)}{\int_0^{+\infty} \int_0^{+\infty} L_\sigma(s,t)\, \pi(s,t)\, ds\, dt}.$$

Typically, Bayes's Rule is not used directly in practice because the formula that it produces for the probability density of m and σ_C given the data and the prior information involves integrals (in the denominator) that cannot be evaluated analytically, and that may be impracticable to compute numerically. Other tools have to be employed to coax the wheels of the Bayesian machinery to turn.

An invention dating back to the 1950s, Markov Chain Monte Carlo (MCMC) sampling,[134] coupled with the availability of fast personal computers, has revolutionized the practice of Bayesian statistics. MCMC frees users from constraints of mathematical tractability, and allows them to employ realistically appropriate Bayesian models and still be able to draw samples from the posterior distribution without computing its density explicitly (for example, q_σ above).

[134] C. Robert and G. Casella. A short history of Markov Chain Monte Carlo: Subjective recollections from incomplete data. *Statistical Science*, 26(1):102–115, 2011. doi:10.1214/10-STS3510

Markov Chain Monte Carlo is an iterative procedure. At each step, first it generates proposed values for the parameters by making random draws from a suitable (generally multivariate) distribution (fittingly called the *proposal distribution*). Then it compares the proposed values with the values that had been accepted in the previous step by computing the ratio of the value that the posterior probability density (Page 159) takes at the proposed parameter values, to the value it takes at the previously accepted parameter values. To compute this

The Russian mathematician Andrey Markov (1856-1922) found that the sequence of consonants and vowels in Alexander Pushkin's *Eugene Onegin* could be described as a random sequence with a particular structure: the probability of the appearance of a vowel or consonant largely depends only on the type of letter immediately preceding it. This model is still in use today to help identify the authors of texts of unknown authorship [Khmelev and Tweedie, 2001].

ratio, α, only the numerator of Bayes's formula needs to be evaluated, not the denominator, which usually is the challenging or impracticable piece to compute.

When the proposal distribution is asymmetrical (as it will be in this case) this ratio is adjusted by a multiplier ρ defined in the R code below (where it is called `rho`). The adjusted ratio is denoted α in the text, and `alpha` in the R code.

When $\alpha > 1$, the proposed values of the parameters are accepted in the current step without further ado. When $\alpha \leqslant 1$, a number z is drawn that is distributed uniformly between 0 and 1: if $z < \alpha$, then the proposed values are still accepted; otherwise, the proposal is rejected and the values of the parameters in the previous step are taken also for the current step.

Since the result of each step depends only on the result of the previous step, the resulting sequence of parameter values is a Markov chain on the space of parameter values. The manner, specified above, of transitioning from one step to the next, ensures that the stationary (or, equilibrium) distribution of this Markov chain is the posterior probability distribution sought.

The chain eventually "forgets" its initial state — which is an arbitrary assignment of values to the parameters —, and the sequence of accepted values of the parameters is like a sample from the posterior distribution, albeit with some dependence that can be alleviated subsequently by *thinning*: for example, by keeping only the value that the Markov chain takes at every 10th or 25th step, say.

Nowadays there are many different ways of implementing MCMC. The procedure sketched above is one of the oldest, called the *Metropolis-Hastings Algorithm* [Metropolis and Ulam, 1949] [Hastings, 1970].[135]

[135] C. P. Robert and G. Casella. *Introducing Monte Carlo Methods with R*. Springer, New York, NY, 2010. ISBN 978-1-4419-1575-7. doi:10.1007/978-1-4419-1576-4

The following R code shows an example of how the Metropolis-Hastings version of Markov Chain Monte Carlo can be used to sample the joint posterior distribution of the Weibull modulus, m, and characteristic strength, σ_C, to produce Bayesian counterparts of the maximum likelihood (Page 193) estimates that were computed above for 30 observations of the rupture stress of alumina coupons modeled as a sample from a Weibull distribution (Page 171).

We begin by defining an R function that computes the logarithm of the numerator of Bayes's Rule.

```
lup = function (theta, x) {
  m = theta[1]; sigmaC = theta[2]
  ## Logarithm of prior density for m
  logprior.m = dnorm(m, mean=8.8, sd=1.25, log=TRUE)
  ## Logarithm of prior density for sigmaC
  logprior.s = dnorm(sigmaC, mean=467, sd=11, log=TRUE)
  ## Log-likelihood function
  loglik = sum(dweibull(x, shape=m, scale=sigmaC, log=TRUE))
  ## The logarithm of the numerator in Bayes's Rule
  ## is the sum of the logarithms of the prior densities
  ## and of the log-likelihood
  return(logprior.m + logprior.s + loglik)
}
```

The R function lup evaluates the logarithm of the numerator of Bayes's rule, $\ln(p_{m,\sigma_C}) + \ln(\pi)$, which is all that is needed to be able to do MCMC. (The name "lup" refers to the logarithm of the unnormalized posterior density, which is the numerator of Bayes's Rule.)

Next we place the determinations of rupture stress that we used above, when discussing maximum likelihood estimation (Page 191), into the vector sigma, and set the stage for MCMC.

```
## Determinations of rupture stress of alumina coupons (MPa)
sigma = c(307, 371, 380, 393, 393, 402, 407, 409,
          411, 428, 430, 434, 435, 437, 441, 445,
          445, 449, 455, 462, 465, 466, 480, 485,
          486, 499, 499, 500, 543, 562)

require(truncnorm)
cv = 0.0485   ## coefficient of variation
K = 1e6       ## Number of steps for the Markov chain

## Coordinates that the Markov chain visits as it moves
## over the possible values for mu and sigmaC
mcmc = array(dim=c(K,2))

## Starting location for the Markov chain
mcmc[1,] = c(m=9, sigmaC=470)

## Counter of the number of times a proposal is accepted
nAccept = 0
```

Finally, we take K steps of the Markov chain defined above, drawing candidate values for the parameters from Gaussian distributions truncated at zero.

Such truncated distribution produces only positive values, which are the only acceptable values for the shape and scale of a Weibull distribution.

```
for (k in 2:K) {
   ## Generate new proposal values for the parameters in the
   ## vicinity of the previous values
   proposal = rtruncnorm(2, a=0, mean=mcmc[k-1,1],
                         sd=cv*mcmc[k-1,2])

   ## Calculate acceptance ratio alpha, corrected using rho
   ## because the proposal distribution is asymmetric
   a1 = dtruncnorm(mcmc[k-1,], a=0, mean=proposal[1],
                   sd=cv*proposal[2])
   a2 = dtruncnorm(proposal, a=0, mean=mcmc[k-1,],
                   sd=cv*mcmc[k-1,])
   rho = a1/a2
   alpha = rho * exp(lup(proposal, x=sigma))/
           exp(lup(mcmc[k-1,], x=sigma))

   if ((alpha > 1) || (runif(1) < alpha)) {
     ## Accept proposed values if a number drawn uniformly
     ## at random from [0,1] is smaller than alpha
     nAccept = nAccept + 1
     mcmc[k,] = proposal } else { mcmc[k,] = mcmc[k-1,] }
}
nAccept/K ## Acceptance rate
```

Once these K steps are completed, we discard the initial 25 % of the chain to remove any memory of the starting values, and keep only every 20th pair of parameter values thereafter to reduce the impact that correlations between accepted values may have upon the estimates of standard uncertainty for the Bayes estimates that we will derive from the MCMC sample.

```
## Discard the initial 25 percent of the chain,
## and keep only every 25th of the remaining values
mcmc = mcmc[seq(0.25*nrow(mcmc), nrow(mcmc), by=25),]
m.TILDE = mcmc[,1]
sigmaC.TILDE = mcmc[,2]
eta.TILDE = sigmaC.TILDE*gamma(1 + 1/m.TILDE)
```

What do we do with such sample? The sky is the limit, really, because by making this sample very large (which can be done at the expense of very quick computation), we characterize it sufficiently well to be able to compute any function of it that will be required, and to do so with high accuracy.

In the case we are considering, this sample comprises pairs of values of m and σ_C (which, *a posteriori*, are no longer independent, because they draw information from the same data). The first thing we do with this sample of pairs of values of the parameters is to compute a value of η from each of these pairs, thus producing a sample from the distribution of the measurand.

```
eta.TILDE = sigmaC.TILDE * gamma(1 + 1/m.TILDE)
```

Then we can summarize this sample in any way we wish: by computing its mean or its median, its standard deviation, coverage intervals of any probability, etc.

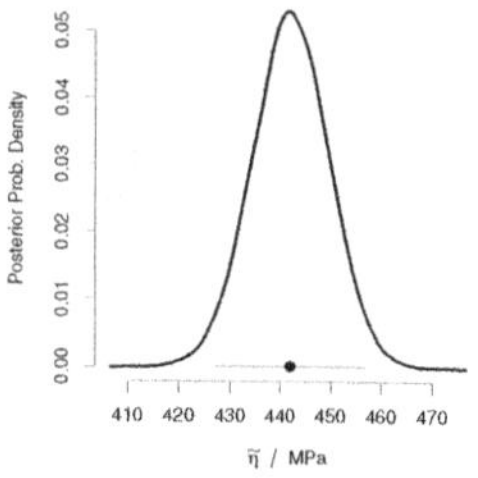

Posterior probability density (Page 159) of η obtained using the simple implementation of the MCMC sampler described above, posterior mean (dot), and 95 % credible interval centered at the posterior mean (line segment).

```
mean(eta.TILDE); sd(eta.TILDE)
quantile(eta.TILDE, probs = c(0.025, 0.975))
```

The MLE and Bayes estimates of η, 443 MPa and 442 MPa, are almost identical, but the associated uncertainties are markedly different: MLE's is 10.4 MPa, while its Bayesian counterpart is 7.5 MPa.

The estimates are almost identical because the information in the data is in very close agreement with the prior information, and because there is enough data to weigh fairly heavily upon the specified prior information.

The uncertainty for the Bayes estimate is appreciably smaller than for the MLE because the prior information is very specific, which the MLE is not privy to. In fact, the MLE may be interpreted as a particular Bayesian estimate (the so-called maximum *a posteriori* estimate) when the parameters are uniformly distributed *a priori* over their ranges, even though using such uniform distribution (Page 167) as a prior distribution in this case is a questionable proposition because both the shape and scale parameters can take any positive values, which form an interval of infinite length.

The power of Bayesian methods lies in the fact that they allow us to incorporate relevant information that the likelihood function may be unable to accommodate. For example, preexisting knowledge about the values of the parameters, or constraints that the parameters must satisfy.

THE MASS FRACTION OF NITRITE IONS in a sample of seawater was measured using Griess's method,[136] based on four determinations obtained under conditions of repeatability:

[136] P. Griess. Bemerkungen zu der abhandlung der hh. weselsky und benedikt "ueber einige azoverbindungen". *Berichte der Deutschen Chemischen Gesellschaft*, 12(1):426–428, 1879. doi:10.1002/cber.187901201117

$$w(\mathrm{NO_2^-}) = 0.1514,\ 0.1523,\ 0.1545,\ 0.1531\ \mathrm{mg/kg}$$

While we might not have any strong prior information about the nitrite levels in this seawater sample, based on the performance of the measurement method we do expect that the relative measurement uncertainty is 1 % to within a factor of 3. We can model this prior knowledge about the standard deviation, σ, of the measurement errors affecting the individual determinations, using a gamma distribution (Page 170) whose 10th and 90th percentiles are 0.33 % and 3 % of 0.150 mg/kg, respectively. Using R we can obtain the parameters of the gamma distribution that has these percentiles as follows:

```
require(rriskDistributions)
get.gamma.par(p = c(0.10, 0.90), q = 0.150*c(1/3, 3)/100)
```

This yields shape $\alpha = 1.696$ and rate $\lambda = 762.3\ \mathrm{kg/mg}$. The following Stan and R codes fit the simple statistical model

$$w_i(\mathrm{NO_2^-}) = \omega + \varepsilon_i$$

to the replicate determinations $i = 1, 2, 3, 4$, where ω denotes the true value of the mass fraction of nitrite in the sample of seawater, and the measurement errors $\{\varepsilon_i\}$ are assumed to be a sample from a Gaussian distribution with mean 0 and standard deviation σ.

The prior information about σ is encapsulated in the gamma distribution specified above. For ω we adopt a weakly informative Gaussian prior distribution.

```
require(rstan)
w = c(0.1514, 0.1523, 0.1545, 0.1531)
m = "data {
      real w[4];
    }
    parameters {
      real<lower=0> omega;
      real<lower=0> sigma;
    }
    model {
      // Prior for true mean mass fraction of nitrite
      omega ~ normal(0, 1);
      // Prior for std. deviation of measurement errors
      sigma ~ gamma(1.696, 762.3);
      // Likelihood
      w ~ normal(omega, sigma);
    }"
fit = stan(model_code = m, data = list(w=w),
           warmup=75000, iter=750000,
           chains=4, cores=4, thin=25)
print(fit, digits=5)
```

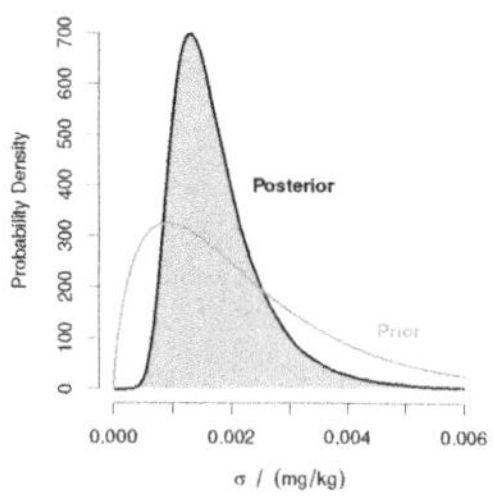

Prior and posterior probability densities for σ. The relative prior uncertainty about σ, which is 77 %, is reduced to 47 % after incorporation of the observations.

The posterior mean of ω is 0.1528 mg/kg, with standard uncertainty 0.0010 mg/kg, which is 50 % larger than the conventional Type A evaluation of the standard uncertainty for the average of the replicates.

Bibliography

A. Agresti. *Analysis of Ordinal Categorical Data*. John Wiley & Sons, Hoboken, NJ, 2nd edition, 2010. ISBN 978-0-470-08289-8. doi:10.1002/9780470594001.

A. Agresti. *An Introduction to Categorical Data Analysis*. John Wiley & Sons, Hoboken, NJ, 3rd edition, 2019. ISBN 978-1119405269.

C. Ainsworth. Sex redefined. *Nature*, 518:288–291, February 2015. doi:10.1038/518288a. News Feature.

D. G. Altman and J. M. Bland. Measurement in medicine: the analysis of method comparison studies. *Journal of the Royal Statistical Society, Series D (The Statistician)*, 32(3):307–317, September 1983. doi:10.2307/2987937.

D. G. Altman and J. M. Bland. Statistics Notes: Diagnostic tests 1: sensitivity and specificity. *British Medical Journal*, 308(6943):1552, 1994. doi:10.1136/bmj.308.6943.1552.

P. W. Anderson. More is different: Broken symmetry and the nature of the hierarchical structure of science. *Science*, 177:393–396, 1972. doi:10.1126/science.177.4047.393.

T. W. Anderson and D. A. Darling. Asymptotic theory of certain "goodness-of-fit" criteria based on stochastic processes. *Annals of Mathematical Statistics*, 23:193–212, 1952. doi:10.1214/aoms/1177729437.

T. M. Apostol. Zeta and related functions. In F. W. J. Olver, D. W. Lozier, R. F. Boisvert, and C. W. Clark, editors, *NIST Handbook of Mathematical Functions*, chapter 25. Cambridge University Press, Cambridge, UK, 2010. ISBN 978-0-521-19225-5.

ASTM. *ASTM E74-13a, Practice of Calibration of Force-Measuring Instruments for Verifying the Force Indication of Testing Machines*. ASTM International, West Conshohocken, PA, 2013. doi:10.1520/E0074-13A.

G. Audi, F.G. Kondev, M. Wang, W. J. Huang, and S. Naimi. The NUBASE2016 evaluation of nuclear properties. *Chinese Physics C*, 41(3):030001–1–138, March 2017. doi:10.1088/1674-1137/41/3/030001.

A. Azzalini and A. Capitanio. *The Skew-Normal and Related Families*. Cambridge University Press, Cambridge, UK, 2014. ISBN 978-1-107-02927-9. doi:10.1017/cbo9781139248891.

A. Baddeley and R. Turner. spatstat: An R package for analyzing spatial point patterns. *Journal of Statistical Software*, 12:1–42, 2005. URL `www.jstatsoft.org/v12/i06/`.

R. Badertscher, T. Berger, and R. Kuhn. Densitometric determination of the fat content of milk and milk products. *International Dairy Journal*, 17(1):20–23, 2007. doi:10.1016/j.idairyj.2005.12.013.

R. E. Barlow and T. Z. Irony. Foundations of statistical quality control. In M. Ghosh and P. K. Pathak, editors, *Current issues in statistical inference: Essays in honor of D. Basu*, volume 17 of *IMS Lecture Notes – Monograph Series*, pages 99–112. Institute of Mathematical Statistics, 1992. ISBN 0-940600-24-2. doi:10.1214/lnms/1215458841.

T. Bartel. Uncertainty in NIST force measurements. *Journal of Research of the National Institute of Standards and Technology*, 110(6):589–603, 2005.

T. Bartel, S. Stoudt, and A. Possolo. Force calibration using errors-in-variables regression and Monte Carlo uncertainty evaluation. *Metrologia*, 53(3):965–980, 2016. doi:10.1088/0026-1394/53/3/965.

E. J. Baxter and B. D. Sherwin. Determining the Hubble constant without the sound horizon scale: measurements from CMB lensing. *Monthly Notices of the Royal Astronomical Society*, 501(2):1823–1835, 2020. doi:10.1093/mnras/staa3706.

Mr. Bayes and Mr. Price. An essay towards solving a problem in the doctrine of chances. *Philosophical Transactions of the Royal Society of London*, 53:370–418, January 1763. doi:10.1098/rstl.1763.0053. Communicated by Mr. Price, in a letter to John Canton.

S. Bell. *A Beginner's Guide to Uncertainty of Measurement*, volume 11 (Issue 2) of *Measurement Good Practice Guide*. National Physical Laboratory, Teddington, Middlesex, United Kingdom, 1999. URL `www.npl.co.uk/publications/guides/a-beginners-guide-to-uncertainty-of-measurement`. Amendments March 2001.

J. M. Bernardo. The concept of exchangeability and its applications. *Far East Journal of Mathematical Sciences (Bhattacharya Memorial Volume)*, 1:111—121, 1996.

K. J. Berry, J. E. Johnston, and Jr. P. W. Mielke. *A Primer of Permutation Statistical Methods*. Springer, Cham, Switzerland, 2019. ISBN 978-3-030-20932-2. doi:10.1007/978-3-030-20933-9.

R. P. Binzel. Pluto-Charon mutual events. *Geophysical Research Letters*, 16(11):1205–1208, November 1989. doi:10.1029/gl016i011p01205.

R. T. Birge. The calculation of errors by the method of least squares. *Physical Review*, 40:207–227, April 1932. doi:10.1103/PhysRev.40.207.

S. Birrer et al. H0LiCOW – IX. Cosmographic analysis of the doubly imaged quasar SDSS 1206+4332 and a new measurement of the Hubble constant. *Monthly Notices of the Royal Astronomical Society*, 484(4):4726–4753, January 2019. doi:10.1093/mnras/stz200.

O. N. Bjørnstad. *Epidemics — Models and Data using R*. Springer, Cham, Switzerland, 2018. ISBN 978-3-319-97486-6. doi:10.1007/978-3-319-97487-3.

M. Blaauw. Methods and code for 'classical' age-modelling of radiocarbon sequences. *Quaternary Geochronology*, 5:512–518, 2010. doi:10.1016/j.quageo.2010.01.002.

M. Blaauw. *clam: Classical Age-Depth Modelling of Cores from Deposits*, 2021. URL `https://CRAN.R-project.org/package=clam`. R package version 2.4.0.

J. M. Bland and D. G. Altman. Statistical methods for assessing agreement between two methods of clinical measurement. *Lancet*, 327:307–310, 1986. doi:10.1016/S0140-6736(86)90837-8.

J. M. Bland and D. G. Altman. Measuring agreement in method comparison studies. *Statistical Methods in Medical Research*, 8:135–160, 1999. doi:10.1177/096228029900800204.

BMJ News and Notes. Epidemiology — Influenza in a boarding school. *British Medical Journal*, 1:586–590, March 1978. doi:10.1136/bmj.1.6112.586.

B. Bolker and R Development Core Team. *bbmle: Tools for General Maximum Likelihood Estimation*, 2020. URL `https://CRAN.R-project.org/package=bbmle`. R package version 1.0.23.1.

G. E. P. Box, G. M. Jenkins, and G. C. Reinsel. *Time Series Analysis: Forecasting and Control*. John Wiley & Sons, Hoboken, NJ, 4th edition, 2008.

B. Braden. The surveyor's area formula. *The College Mathematics Journal*, 17(4):326–337, 1986. doi:10.2307/2686282.

W. T. Brande. *Dictionary of Science, Literature, and Art: Comprising the History, Description, and Scientific Principles of Every Branch of Human Knowledge*. Longman, Brown, Green, and Longmans, London, UK, 1842.

A. Bredius. A new Vermeer. *The Burlington Magazine for Connoisseurs*, 71(416):210–211, November 1937.

C. D. Brown and H. T. Davis. Receiver operating characteristics curves and related decision measures: a tutorial. *Chemometrics and Intelligent Laboratory Systems*, 80 (1):24–38, 2006. doi:10.1016/j.chemolab.2005.05.004.

R. L. Burger, L. C. Salazar, J. Nesbitt, E. Washburn, and L. Fehren-Schmitz. New AMS dates for Machu Picchu: results and implications. *Antiquity*, 95(383):1265–1279, 2021. doi:10.15184/aqy.2021.99.

H. Burgess and B. Spangler. Consensus building. In G. Burgess and H. Burgess, editors, *Beyond Intractability*. Conflict Research Consortium, University of Colorado, Boulder, Colorado, USA, September 2003. URL `www.beyondintractability.org/essay/consensus-building`.

P.-C. Bürkner. Advanced Bayesian multilevel modeling with the R package brms. *The R Journal*, 10(1):395–411, 2018. doi:10.32614/RJ-2018-017.

K. P. Burnham and D. R. Anderson. Multimodel inference: Understanding AIC and BIC in model selection. *Sociological Methods & Research*, 33(2):261–304, November 2004. doi:10.1177/0049124104268644.

S. G. Burrard. Mount Everest: The story of a long controversy. *Nature*, 71:42–46, November 1904. doi:10.1038/071042a0.

N. Campione and D. C. Evans. A universal scaling relationship between body mass and proximal limb bone dimensions in quadrupedal terrestrial tetrapods. *BMC Biology*, 10:60, 2012. doi:10.1186/1741-7007-10-60.

N. E. Campione and D. C. Evans. The accuracy and precision of body mass estimation in non-avian dinosaurs. *Biological Reviews*, 95(6):1759–1797, 2020. doi:10.1111/brv.12638.

A. Canty and B. Ripley. *boot: Bootstrap R (S-Plus) Functions*, 2020. URL `cran.r-project.org/web/packages/boot/`. R package version 1.3-25.

A. Canty and B. D. Ripley. *boot: Bootstrap R (S-Plus) Functions*, 2021. URL `cran.r-project.org/web/packages/boot/`. R package version 1.3-28.

B. Carpenter, A. Gelman, M. Hoffman, D. Lee, B. Goodrich, M. Betancourt, M. Brubaker, J. Guo, P. Li, and A. Riddell. Stan: A probabilistic programming language. *Journal of Statistical Software*, 76(1):1–32, 2017. doi:10.18637/jss.v076.i01.

R. J. Carroll, D. Ruppert, L. A. Stefanski, and C. M. Crainiceanu. *Measurement Error in Nonlinear Models — A Modern Perspective*. Chapman & Hall/CRC, Boca Raton, Florida, second edition, 2006.

B. Carstensen. *Comparing Clinical Measurement Methods*. John Wiley & Sons, Chichester, UK, 2010.

B. Carstensen, L. Gurrin, C. T. Ekstrøm, and M. Figurski. *MethComp: Analysis of Agreement in Method Comparison Studies*, 2020. URL `https://CRAN.R-project.org/package=MethComp`. R package version 1.30.0.

J. Chambers, W. Cleveland, B. Kleiner, and P. Tukey. *Graphical Methods for Data Analysis*. Wadsworth, Belmont, CA, 1983.

F. Chayes. On correlation between variables of constant sum. *Journal of Geophysical Research*, 65(12):4185–4193, 1960. doi:10.1029/JZ065i012p04185.

F. Chayes. *Ratio Correlation: A Manual for Students of Petrology and Geochemistry*. University of Chicago Press, Chicago, Illinois, 1971.

G. C.-F. Chen et al. A SHARP view of H0LiCOW: H_0 from three time-delay gravitational lens systems with adaptive optics imaging. *Monthly Notices of the Royal Astronomical Society*, 490(2):1743–1773, September 2019. doi:10.1093/mnras/stz2547.

M. R. Chernick. *Bootstrap Methods: A Guide for Practitioners and Researchers*. John Wiley & Sons, Hoboken, NJ, second edition, 2008. ISBN 978-0-471-75621-7.

H. Chipman, E. I. George, R. B. Gramacy, and R. McCulloch. Bayesian treed response surface models. *Wiley Interdisciplinary Reviews: Data Mining and Knowledge Discovery*, 3(4):298–305, 2013. doi:10.1002/widm.1094.

Y. Cho, L. Hu, H. Hou, et al. The tiger genome and comparative analysis with lion and snow leopard genomes. *Nature Communications*, 4:2433, September 2013. doi:10.1038/ncomms3433.

R. H. B. Christensen. ordinal — Regression Models for Ordinal Data, 2019. R package version 2019.12-10.

G. Clark, A. Gonye, and S. J. Miller. Lessons from the German Tank Problem. *arXiv e-prints*, page arXiv:2101.08162 [stat.OT], 2021. URL `https://arxiv.org/abs/1905.12362`.

W. S. Cleveland, E. Grosse, and W. M. Shyu. Local regression models. In J. M. Chambers and T. J. Hastie, editors, *Statistical Models in S*, chapter 8. Wadsworth & Brooks/Cole, Pacific Grove, California, 1992.

W. G. Cochran. The combination of estimates from different experiments. *Biometrics*, 10(1):101–129, March 1954. doi:10.2307/3001666.

Consultative Committee for Thermometry. *Guide to the Realization of the ITS-90*. Bureau International des Poids et Mesures (BIPM), Sèvres, France, 2018. URL `https://www.bipm.org/en/committees/cc/cct/guide-its90.html`.

S. Cowley and J. Silver-Greenberg. These Machines Can Put You in Jail. Don't Trust Them. *The New York Times*, November 3, 2019. Business Section.

P. E. Damon, D. J. Donahue, B. H. Gore, A. L. Hatheway, A. J. T. Jull, T. W. Linick, P. J. Sercel, L. J. Toolin, C. R. Bronk, E. T. Hall, R. E. M. Hedges, R. Housley, I. A. Law, C. Perry, G. Bonani, S. Trumbore, W. Woelfli, J. C. Ambers, S. G. E. Bowman, M. N. Leese, and M. S. Tite. Radiocarbon dating of the Shroud of Turin. *Nature*, 337:611–615, February 1989. doi:10.1038/337611a0.

J. Damuth. Interspecific allometry of population density in mammals and other animals: the independence of body mass and population energy-use. *Biological Journal of the Linnean Society*, 31:193–246, 1987. doi:10.1111/j.1095-8312.1987.tb01990.x.

J. Damuth. A macroevolutionary explanation for energy equivalence in the scaling of body size and population density. *The American Naturalist*, 169(5):621–631, 2007. doi:10.1086/513495. Associate Editor: Claire de Mazancourt and Editor: Donald L. DeAngelis.

A. C. Davison and D. Hinkley. *Bootstrap Methods and their Applications*. Cambridge University Press, Cambridge, UK, 1997. ISBN 0-521-57471-4. URL `statwww.epfl.ch/davison/BMA/`.

B. de Finetti. Funzione caratteristica di un fenomeno aleatorio. *Atti della R. Academia Nazionale dei Lincei*,

Serie 6. Memorie, Classe di Scienze Fisiche, Mathematice e Naturale, 4:251–299, 1930.

T. de Jaeger, B. E. Stahl, W. Zheng, A. V. Filippenko, A. G. Riess, and L. Galbany. A measurement of the Hubble constant from Type II supernovae. *Monthly Notices of the Royal Astronomical Society*, 496(3):3402–3411, 2020. doi:10.1093/mnras/staa1801.

R. Dedekind. *The nature and meaning of numbers*. Open Court Publishing Company, Chicago, 1901. Translated from the German by W. W. Beman.

M. H. DeGroot and M. J. Schervish. *Probability and Statistics*. Addison-Wesley, Boston, MA, 4th edition, 2012.

W. E. Deming. *Statistical Adjustment of Data*. John Wiley & Sons, New York, NY, 1943.

P. Denzel, J. P. Coles, P. Saha, and L. L. R. Williams. The Hubble constant from eight time-delay galaxy lenses. *Monthly Notices of the Royal Astronomical Society*, 501 (1):784–801, 2020. doi:10.1093/mnras/staa3603.

P. Diaconis and B. Efron. Computer-intensive methods in statistics. *Scientific American*, 248:116–130, 1983.

A. E. Dolbear. The cricket as a thermometer. *The American Naturalist*, 31(371):970–971, 1897. doi:10.2307/2453256.

A. Domínguez, R. Wojtak, J. Finke, M. Ajello, K. Helgason, F. Prada, A. Desai, V. Paliya, L. Marcotulli, and D. H. Hartmann. A new measurement of the Hubble Constant and matter content of the universe using extragalactic background light γ-ray attenuation. *The Astrophysical Journal*, 885(2):137, November 2019. doi:10.3847/1538-4357/ab4a0e.

R. L. Duncombe and P. K. Seidelmann. A history of the determination of Pluto's mass. *Icarus*, 44:12–18, 1980. doi:10.1016/0019-1035(80)90048-2.

K. Dutta, A. Roy, Ruchika, A. A. Sen, and M. M. Sheikh-Jabbari. Cosmology with low-redshift observations: No signal for new physics. *Physical Review D*, 100:103501, November 2019. doi:10.1103/PhysRevD.100.103501.

B. Efron and R. J. Tibshirani. *An Introduction to the Bootstrap*. Chapman & Hall, London, UK, 1993.

J. Elío, G. Cinelli, P. Bossew, J. L. Gutiérrez-Villanueva, T. Tollefsen, M. De Cort, A. Nogarotto, and R. Braga. The first version of the Pan-European Indoor Radon Map. *Natural Hazards and Earth System Sciences*, 19(11): 2451–2464, 2019. doi:10.5194/nhess-19-2451-2019.

D. R. Ferguson. Construction of curves and surfaces using numerical optimization techniques. *Computer-Aided Design*, 18(1):15–21, 1986. doi:10.1016/S0010-4485(86)80004-5.

R. A. Fisher. *Statistical Methods for Research Workers*. Hafner Publishing Company, New York, NY, 14th edition, 1973.

M. A. Fligner and T. J. Killeen. Distribution-free two-sample tests for scale. *Journal of the American Statistical Association*, 71(353):210–213, March 1976. doi:10.2307/2285771.

D. Freedman, R. Pisani, and R. Purves. *Statistics*. W. W. Norton & Company, New York, NY, 4th edition, 2007. ISBN 978-0-393-92972-0.

W. L. Freedman. Measurements of the Hubble Constant: Tensions in perspective. *The Astrophysical Journal*, 919 (1):16, 2021. doi:10.3847/1538-4357/ac0e95.

W. L. Freedman et al. The Carnegie-Chicago Hubble Program. VIII. An independent determination of the Hubble Constant based on the Tip of the Red Giant Branch. *The Astrophysical Journal*, 882(1):34, August 2019. doi:10.3847/1538-4357/ab2f73.

L. M. Friedman, C. D. Furberg, D. DeMets, D. M. Reboussin, and C. B. Granger. *Fundamentals of Clinical Trials*. Springer, Switzerland, 5th edition, 2015.

H. Frings and M. Frings. The effects of temperature on chirp-rate of male cone-headed grasshoppers, *Neoconocephalus ensiger*. *Journal of Experimental Zoology*, 134:411–425, 1957. doi:10.1002/jez.1401340302.

X. Fuentes-Arderiu and D. Dot-Bach. Measurement uncertainty in manual differential leukocyte counting. *Clinical Chemistry and Laboratory Medicine*, 47(1):112–115, 2009. doi:10.1515/LM.2009.014.

X. Fuentes-Arderiu, M. García-Panyella, and D. Dot-Bach. Between-examiner reproducibility in manual differential leukocyte counting. *Accreditation and Quality Assurance*, 12:643–645, 2007. doi:10.1007/s00769-007-0323-0.

C. F. Gauss. Summarische Uberficht der zur bestimmung der bahnen der beyden neuen hauptplaneten augewanden methoden. *Monatliche Correspondenz zur Beförderung der Erd- und Himmels- Kunde*, XX(Part B, July-December, Section XVII):197–224, September 1809.

C. F. Gauss. Theoria combinationis observationum erroribus minimis obnoxiae. In *Werke, Band IV, Wahrscheinlichkeitsrechnung und Geometrie*. Könighlichen Gesellschaft der Wissenschaften, Göttingen, 1823. URL http://gdz.sub.uni-goettingen.de.

C. F. Gauss and G. W. Stewart. *Theory of the Combination of Observations Least Subject to Errors*. Classics in Applied Mathematics. SIAM (Society for Industrial and Applied Mathematics), Philadelphia, 1995. ISBN 978-0-89871-347-3. doi:10.1137/1.9781611971248.

R. Geary. Comparison of the concepts of efficiency and closeness for consistent estimates of a parameter. *Biometrika*, 33:123–128, 1944. doi:10.2307/2334111.

A. Gelman. Prior distributions for variance parameters in hierarchical models (comment on article by Browne and Draper). *Bayesian Analysis*, 1(3):515–533, 2006. doi:10.1214/06-BA117A.

A. Gelman, J. B. Carlin, H. S. Stern, and D. B. Rubin. *Bayesian Data Analysis*. Chapman & Hall / CRC, Boca Raton, FL, 2nd edition, 2003.

S. Geman and D. Geman. Stochastic relaxation, Gibbs distributions, and the Bayesian restoration of images. *IEEE Transactions on Pattern Analysis and Machine Intelligence*, 6:721–741, 1984.

A. Ghalanos and S. Theussl. *Rsolnp: General Nonlinear Optimization Using Augmented Lagrange Multiplier Method*, 2015. R package version 1.16.

G. Giordano, F. Blanchini, R. Bruno, P. Colaneri, A. Di Filippo, A. Di Matteo, and M. Colaneri. Modelling the COVID-19 epidemic and implementation of population-wide interventions in Italy. *Nature Medicine*, 26:855–860, June 2020. doi:10.1038/s41591-020-0883-7.

J. D. Giorgini. Status of the JPL Horizons Ephemeris System. In *IAU General Assembly*, volume 29, page 2256293, August 2015. URL `https://ssd.jpl.nasa.gov/`.

P. G. Gottschalk and J. R. Dunn. The five-parameter logistic: A characterization and comparison with the four-parameter logistic. *Analytical Biochemistry*, pages 54–65, 2005. doi:10.1016/j.ab.2005.04.035.

R. B. Gramacy. tgp: An R package for Bayesian nonstationary, semiparametric nonlinear regression and design by treed Gaussian process models. *Journal of Statistical Software*, 19(9):1–46, 2007. URL `www.jstatsoft.org/v19/i09`.

R. B. Gramacy. *Surrogates: Gaussian Process Modeling, Design, and Optimization for the Applied Sciences*. CRC Press, Boca Raton, FL, 2020. ISBN 978-0-367-41542-6.

R. B. Gramacy and H. K. H. Lee. Bayesian treed Gaussian process models with an application to computer modeling. *Journal of the American Statistical Association*, 103:1119–1130, 2008.

P. Griess. Bemerkungen zu der abhandlung der hh. weselsky und benedikt "ueber einige azoverbindungen". *Berichte der Deutschen Chemischen Gesellschaft*, 12(1): 426–428, 1879. doi:10.1002/cber.187901201117.

G. Grimmett and D. Welsh. *Probability: An Introduction*. Oxford University Press, Oxford, UK, 2nd edition, 2014. ISBN 978-0-19-870997-8.

F. M. Guerra, S. Bolotin, G. Lim, J. Heffernan, S. L. Deeks, Y. Li, and N. S. Crowcroft. The basic reproduction number (R_0) of measles: a systematic review. *The Lancet Infectious Diseases*, 17:e420–e428, 2017. doi:10.1016/s1473-3099(17)30307-9.

R. W. Gurney and E. U. Condon. Wave mechanics and radioactive disintegration. *Nature*, 122:439, September 1928. doi:10.1038/122439a0.

K. Gurung. *Fractal Dimension in Architecture: An Exploration of Spatial Dimension*. Master thesis, Anhalt University of Applied Sciences, Köthen, Germany, August 2017.

A. Hájek. Dutch Book Arguments. In P. Anand, P. K. Pattanaik, and C. Puppe, editors, *The Handbook of Rational & Social Choice*, chapter 7, pages 173–195. Oxford University Press, Oxford, UK, 2009. ISBN 978-0-19-929042-0.

B. D. Hall and D. R. White. *An Introduction to Measurement Uncertainty*. Measurement Standards Laboratory of New Zealand, Lower Hutt, New Zealand, 2018. ISBN 978-0-473-40581-6. doi:10.5281/zenodo.3872590. URL `https://zenodo.org/record/3872590`. Also available from `www.lulu.com`.

T. Hastie, R. Tibshirani, and J. Friedman. *The Elements of Statistical Learning: Data Mining, Inference, and Prediction*. Springer-Verlag, New York, second edition, 2009. URL `statweb.stanford.edu/~tibs/ElemStatLearn/`.

W. K. Hastings. Monte Carlo sampling methods using Markov chains and their applications. *Biometrika*, 57: 97–109, April 1970. doi:10.1093/biomet/57.1.97.

T. J. Heaton, M. Blaauw, P. G. Blackwell, C. Bronk Ramsey, P. J. Reimer, and E. M. Scott. The INTCAL20 approach to radiocarbon calibration curve construction: a new methodology using Bayesian splines and errors-in-variables. *Radiocarbon*, pages 1–43, 2020. doi:10.1017/RDC.2020.46.

J. M. Heffernan, R. J. Smith, and L. M. Wahl. Perspectives on the basic reproductive ratio. *Journal of The Royal Society Interface*, 2:281–293, 2005. doi:10.1098/rsif.2005.0042.

T. C. Hesterberg. What teachers should know about the bootstrap: Resampling in the undergraduate statistics curriculum. *The American Statistician*, 69(4):371–386, November 2015. doi:10.1080/00031305.2015.1089789.

H. W. Hethcote. The mathematics of infectious diseases. *SIAM Review*, 42:599–653, 2000. doi:10.1137/S0036144500371907.

J. P. T. Higgins, J. Thomas, J. Chandler, M. Cumpston, T. Li, M. J. Page, and V. A. Welch, editors. *Cochrane Handbook for Systematic Reviews of Interventions*. John Wiley & Sons, Hoboken, NJ, second edition, 2019. ISBN 978-1-119-53662-8.

J. L. Hodges and E. L. Lehmann. Estimates of location based on rank tests. *The Annals of Mathematical Statistics*, 34(2):598–611, June 1963. doi:10.1214/aoms/1177704172.

P.G. Hoel, S. C. Port, and C. J. Stone. *Introduction to Stochastic Processes*. Waveland Press, 1972.

V. Hoffmann, M. Kasik, P. K. Robinson, and C. Venzago. Glow discharge mass spectrometry. *Analytical and Bioanalytical Chemistry*, 381:173–188, 2005. doi:10.1007/s00216-004-2933-2.

M. Hollander, D. A. Wolfe, and E. Chicken. *Nonparametric Statistical Methods*. John Wiley & Sons, Hoboken, NJ, 3rd edition, 2014. ISBN 978-0-470-38737-5.

O. Hönigschmid, S. Horovitz, T. W. Richards, and M. E. Lembert. Das Atomgewicht des Urans und des Bleis. *Zeitschrift für analytische Chemie*, 54:70–72, 1915. doi:10.1007/BF01453144.

K. Hotokezaka et al. A Hubble constant measurement from superluminal motion of the jet in GW170817. *Nature Astronomy*, 3:940–944, July 2019. doi:10.1038/s41550-019-0820-1.

E. Hubble. A relation between distance and radial velocity among extra-galactic nebulae. *Proceedings of the National Academy of Sciences*, 15(3):168–173, 1929. doi:10.1073/pnas.15.3.168.

J. S. Hunter. The exponentially weighted moving average. *Journal of Quality Technology*, 18(4):203–210, 1986. doi:10.1080/00224065.1986.11979014.

R. J. Hyndman and G. Athanasopoulos. *Forecasting: Principles and Practice*. OTexts, Melbourne, Australia, second edition, 2018. ISBN 978-0-9875071-1-2. URL `http://OTexts.com/fpp2/`.

R.J. Hyndman and Y. Khandakar. Automatic time series forecasting: the forecast package for R. *Journal of Statistical Software*, 27:1–22, July 2008.

International Organization of Legal Metrology (OIML). *Weights of classes E_1, E_1, E_2, F_1, F_2, M_{1-2}, M_2, M_{2-3}, and M_3 — Part 1: Metrological and technical requirements.* Bureau International de Métrologie Légale (OIML), Paris, France, 2004. URL `https://www.oiml.org/en/files/pdf_r/r111-1-e04.pdf`. International Recommendation OIML R 111-1 Edition 2004 (E).

B. Ivanović, B. Milošević, and M. Obradović. *symmetry: Testing for Symmetry of Data and Model Residuals*, 2020. URL `https://CRAN.R-project.org/package=symmetry`. R package version 0.2.1.

Z. L. Jabbour and S. L. Yaniv. The kilogram and measurements of mass and force. *Journal of Research of the National Institute of Standards and Technology*, 106(1): 25–46, January–February 2001.

O. B. James and E. D. Jackson. Petrology of the Apollo 11 ilmenite basalts. *Journal of Geophysical Research*, 75 (29):5793–5824, 1970. doi:10.1029/JB075i029p05793.

H. Jeffreys. *Theory of Probability*. Oxford University Press, London, 1939.

H. Jeffreys. *Theory of Probability*. Oxford University Press, London, UK, 3rd edition, 1961. Corrected Impression, 1967.

H. Jeffreys. *Scientific Inference*. Cambridge University Press, London, third edition, 1973.

N. L. Johnson, A. W. Kemp, and S. Kotz. *Univariate Discrete Distributions*. John Wiley & Sons, Hoboken, NJ, Third edition, 2005. ISBN 0-471-27246-9.

V. E. Johnson. Revised standards for statistical evidence. *Proceedings of the National Academy of Sciences*, 110: 19313–19317, 2013. doi:10.1073/pnas.1313476110.

W. E. Johnson. *Logic, Part III — The Logical Foundations of Science*. Cambridge University Press, London, UK, 1924.

Joint Committee for Guides in Metrology. *Evaluation of measurement data — Supplement 1 to the "Guide to the expression of uncertainty in measurement" — Propagation of distributions using a Monte Carlo method*. International Bureau of Weights and Measures (BIPM), Sèvres, France, 2008. URL www.bipm.org/en/publications/guides/gum.html. BIPM, IEC, IFCC, ILAC, ISO, IUPAC, IUPAP and OIML, JCGM 101:2008.

Joint Committee for Guides in Metrology. *Guide to the expression of uncertainty in measurement — Part 6: Developing and using measurement models*. International Bureau of Weights and Measures (BIPM), Sèvres, France, 2020. BIPM, IEC, IFCC, ILAC, ISO, IUPAC, IUPAP and OIML, JCGM GUM-6:2020.

Joint Committee for Guides in Metrology (JCGM). *Evaluation of Measurement Data — Guide to the Expression of Uncertainty in Measurement*. International Bureau of Weights and Measures (BIPM), Sèvres, France, 2008. URL www.bipm.org/en/publications/guides/gum.html. BIPM, IEC, IFCC, ILAC, ISO, IUPAC, IUPAP and OIML, JCGM 100:2008, GUM 1995 with minor corrections.

K. E. Jones, J. Bielby, M. Cardillo, S. A. Fritz, J. O'Dell, C. D. L. Orme, K. Safi, W. Sechrest, E. H. Boakes, C. Carbone, C. Connolly, M. J. Cutts, J. K. Foster, R. Grenyer, M. Habib, C. A. Plaster, S. A. Price, E. A. Rigby, J. Rist, A. Teacher, O. R. P. Bininda-Emonds, J. L. Gittleman, G. M. Mace, and A. Purvis. PanTHERIA: a species-level database of life history, ecology, and geography of extant and recently extinct mammals. *Ecology*, 90(9):2648–2648, 2009. doi:10.1890/08-1494.1. Metadata: https://esapubs.org/archive/ecol/E090/184/metadata.htm.

F. Juarez and S. Singh. Incidence of induced abortion by age and state, Mexico, 2009: new estimates using a modified methodology. *International Perspectives on Sexual and Reproductive Health*, 38:58–67, June 2012. doi:10.1363/3805812.

F. Juarez, S. Singh, S. G. Garcia, and C. D. Olavarrieta. Estimates of induced abortion in Mexico: what's changed between 1990 and 2006? *International Family Planning Perspectives*, 34:158–168, 2008. doi:10.1363/ifpp.34.158.08.

A. F. Karr. Poisson process. In S. Kotz, C. B. Read, N. Balakrishnan, B. Vidakovic, and N. L. Johnson, editors, *Encyclopedia of Statistical Sciences*, volume 3, pages 1910–1918. John Wiley & Sons, Hoboken, NJ, second edition, 2006. ISBN 978-0-471-15044-2. doi:10.1002/0471667196.

B. Keisch. Dating works of art through their natural radioactivity: Improvements and applications. *Science*, 160(3826):413–415, 1968. doi:10.1126/science.160.3826.413.

B. Keisch, R. L. Feller, A. S. Levine, and R. R. Edwards. Dating and authenticating works of art by measurement of natural alpha emitters. *Science*, 155(3767): 1238–1242, 1967. doi:10.1126/science.155.3767.1238.

C. Kendall and T. B. Coplen. Distribution of oxygen-18 and deuterium in river waters across the United States. *Hydrological Processes*, 15:1363–1393, 2001. doi:10.1002/hyp.217.

W. O. Kermack and A. G. McKendrick. A contribution to the mathematical theory of epidemics. *Proceedings of the Royal Society of London A: Mathematical, Physical and Engineering Sciences*, 115:700–721, August 1927. doi:10.1098/rspa.1927.0118.

D. V. Khmelev and F. J. Tweedie. Using Markov chains for identification of writers. *Literary and Linguistic Computing*, 16:299–307, 2001. doi:10.1093/llc/16.3.299.

C. Klein and B. Dutrow. *Manual of Mineral Science*. John Wiley & Sons, Hoboken, NJ, 23rd edition, 2007. ISBN 978-0-471-72157-4.

A. Koepke, T. Lafarge, A. Possolo, and B. Toman. Consensus building for interlaboratory studies, key comparisons, and meta-analysis. *Metrologia*, 54(3):S34–S62, 2017. doi:10.1088/1681-7575/aa6c0e.

F.G. Kondev, M. Wang, W.J. Huang, S. Naimi, and G. Audi. The NUBASE2020 evaluation of nuclear physics properties. *Chinese Physics C*, 45(3):030001, March 2021. doi:10.1088/1674-1137/abddae.

K. Köprücü and E. Seker. Acute toxicity of deltamethrin for freshwater mussel, *Unio elongatulus eucirrus* Bourguignat. *Bulletin of Environmental Contamination and Toxicology*, 80:1–4, 2008. doi:10.1007/s00128-007-9254-z.

T. Lafarge and A. Possolo. The NIST Uncertainty Machine. *NCSLI Measure Journal of Measurement Science*, 10(3):20–27, September 2015. doi:10.1080/19315775.2015.11721732.

D. Lara, J. Strickler, C. D. Olavarrieta, and C. Ellertson. Measuring induced abortion in Mexico: A comparison of four methodologies. *Sociological Methods & Research*, 32(4):529–558, May 2004. doi:10.1177/0049124103262685.

I. Lavagnini and F. Magno. A statistical overview on univariate calibration, inverse regression, and detection limits: Application to gas chromatography/mass spectrometry technique. *Mass Spectrometry Reviews*, 26 (1):1–18, 2007. doi:10.1002/mas.20100.

G. Lemaître. Un univers homogène de masse constante et de rayon croissant rendant compte de la vitesse radiale des nébuleuses extra-galactiques. *Annales de la Société Scientifique de Bruxelles A*, 47:49–59, 1927.

G. Lemaître. Republication of: A homogeneous universe of constant mass and increasing radius accounting for the radial velocity of extra-galactic nebulae. *General Relativity and Gravitation*, 45:1635–1646, 2013. doi:10.1007/s10714-013-1548-3.

A. Lindén and S. Mäntyniemi. Using the negative binomial distribution to model overdispersion in ecological count data. *Ecology*, 92(7):1414–1421, 2011. doi:10.1890/10-1831.1.

A. Lucas, G. J. Hudson, P. Simpson, T. J. Cole, and B. A. Baker. An automated enzymic micromethod for the measurement of fat in human milk. *Journal of Dairy Research*, 54:487–492, November 1987. doi:10.1017/S0022029900025693.

E. Lukacs. A characterization of the normal distribution. *Annals of Mathematical Statistics*, 13(1):91–93, March 1942. doi:10.1214/aoms/1177731647.

E. Macaulay et al. First cosmological results using Type Ia supernovae from the Dark Energy Survey: measurement of the Hubble constant. *Monthly Notices of the Royal Astronomical Society*, 486(2):2184–2196, April 2019. doi:10.1093/mnras/stz978.

M. Maechler, P. Rousseeuw, C. Croux, V. Todorov, A. Ruckstuhl, M. Salibian-Barrera, T. Verbeke, M. Koller, E. L. T. Conceição, and M. A. di Palma. *robustbase: Basic Robust Statistics*, 2021. URL `http://CRAN.R-project.org/package=robustbase`. R package version 0.93-9.

B. Mandelbrot. How long is the coast of Britain? Statistical self-similarity and fractional dimension. *Science*, 156:636–638, May 1967. doi:10.1126/science.156.3775.636.

B. W. Mangum and G. T. Furukawa. *Guidelines for Realizing the International Temperature Scale of 1990 (ITS-90)*. National Institute of Standards and Technology, Gaithersburg, MD, 1990. NIST Technical Note 1265.

B. W. Mangum, G. F. Strouse, and W. F. Guthrie. *CCT-K3: Key Comparison of Realizations over the Range 83.8058 K to 933.473 K*. National Institute of Standards and Technology, Gaithersburg, MD, 2002. doi:10.6028/NIST.TN.1450. NIST Technical Note 1450.

E. J. Marsh, M. C. Bruno, S. C. Fritz, P. Baker, J. M. Capriles, and C. A. Hastorf. IntCal, SHCal, or a Mixed Curve? Choosing a ^{14}C calibration curve for archaeological and paleoenvironmental records from tropical South America. *Radiocarbon*, 60(3):925–940, 2018. doi:10.1017/RDC.2018.16.

C. R. Marshall, D. V. Latorre, C. J. Wilson, T. M. Frank, K. M. Magoulick, J. B. Zimmt, and A. W. Poust. Absolute abundance and preservation rate of *Tyrannosaurus rex*. *Science*, 372(6539):284–287, 2021. doi:10.1126/science.abc8300.

M. Martcheva. *An Introduction to Mathematical Epidemiology*, volume 61 of *Texts in Applied Mathematics*. Springer, New York, NY, 2010. ISBN 978-1-4899-7611-6. doi:10.1007/978-1-4899-7612-3.

M.D. Mastrandrea, K.J. Mach, G. Plattner, O. Edenhofer, T. F. Stocker, C. B. Field, K. L. Ebi, and P.R. Matschoss. The IPCC AR5 guidance note on consistent treatment of uncertainties: a common approach across the working groups. *Climatic Change*, 108:675–691, 2011. doi:10.1007/s10584-011-0178-6. Special Issue: Guidance for Characterizing and Communicating Uncertainty and Confidence in the Intergovernmental Panel on Climate Change.

P. McCullagh. Regression models for ordinal data. *Journal of the Royal Statistical Society, Series B (Methodological)*, 42(2):109–142, 1980. doi:10.1111/j.2517-6161.1980.tb01109.x.

P. McCullagh and J. A. Nelder. *Generalized Linear Models*. Chapman & Hall / CRC, London, UK, 2nd edition, 1989.

D. A. McPhee, I. M. Parsonson, A. J. Della-Porta, and R. G. Jarrett. Teratogenicity of Australian Simbu serogroup and some other Bunyaviridae viruses: the embryonated chicken egg as a model. *Infection and Immunity*, 43:413–420, 1984. doi:10.1128/iai.43.1.413-420.1984.

J. Meija. Atomic weights of the elements: From measurements to the periodic table. In M. J.T. Milton, D. S. Wiersma, C. J. Williams, and M. Sega, editors, *New Frontiers for Metrology: From Biology and Chemistry to Quantum and Data Science*, volume 206 of *Proceedings of the International School of Physics "Enrico Fermi"*, pages 77–93. IOS Press, Amsterdam, The Netherlands, 2021. ISBN 978-1-64368-246-4. doi:10.3254/ENFI210019.

J. Meija, B. Methven, S. Tong, O. Mihai, K. Swider, P. Grinberg, Z. Mester, and L. Yang. HIPB-1: High Purity Lead Certified Reference Material for Lead Mass Fraction, Atomic Weight, Isotopic Composition and Elemental Impurities. National Research Council Canada, Ottawa, 2020.

D. Mendeleev. Die periodische Gesetzmässigkeit der Elemente. *Annalen der Chemie und Pharmacie*, VIII. Suplementband:133–229, 1871.

N. Metropolis and S. Ulam. The Monte Carlo Method. *Journal of the American Statistical Association*, 44:335–341, September 1949.

C. Meyer. The Lunar Sample Compendium. `https://curator.jsc.nasa.gov/lunar/lsc/`, 2011. NASA Astromaterials Research & Exploration Science.

W. Miao, Y. R. Gel, and J. L. Gastwirth. A new test of symmetry about an unknown median. In A. C. Hsiung, Z. Ying, and C.-H. Zhang, editors, *Random Walk, Sequential Analysis and Related Topics: A Festschrift in Honor of Yuan-Shih Chow*, pages 199–214. World Scientific Publishing Company, Singapore, 2006. doi:10.1142/9789812772558_0013.

C. Michotte, G. Ratel, S. Courte, K. Kossert, O. Nähle, R. Dersch, T. Branger, C. Bobin, A. Yunoki, and Y. Sato. BIPM comparison BIPM.RI(II)-K1.Fe-59 of activity measurements of the radionuclide ^{59}Fe for the PTB (Germany), LNE-LNHB (France) and the NMIJ (Japan), and the linked APMP.RI(II)-K2.Fe-59 comparison. *Metrologia*, 57(1A):06003, January 2020. doi:10.1088/0026-1394/57/1a/06003.

M. G. Morgan and M. Henrion. *Uncertainty — A Guide to Dealing with Uncertainty in Quantitative Risk and Policy Analysis*. Cambridge University Press, New York, NY, first paperback edition, 1992. 10th printing, 2007.

D. E. Morris, J. E. Oakley, and J. A. Crowe. A web-based tool for eliciting probability distributions from experts. *Environmental Modelling & Software*, 52:1–4, February 2014. doi:10.1016/j.envsoft.2013.10.010.

F. Mosteller and J. W. Tukey. *Data Analysis and Regression*. Addison-Wesley Publishing Company, Reading, Massachusetts, 1977. ISBN 0-201-04854-X.

F. Mosteller and J. W. Tukey. Data Analysis, including Statistics. In *The Collected Works of John W. Tukey*, volume IV: Philosophy and Principles of Data Analysis: 1965-1986, chapter 15, pages 601–720. Wadsworth & Brooks Cole, Monterey, CA, 1986. ISBN 0-534-05101-4.

S. Mukherjee et al. First measurement of the Hubble parameter from bright binary black hole GW190521, 2020.

J. W. Munch. *Method 524.2. Measurement of Purgeable Organic Compounds in Water by Capillary Column Gas Chromatography/Mass Spectrometry.* National Exposure Research Laboratory, Office of Research and Development, U.S. Environmental Protection Agency, Cincinnati, OH, 1995. Revision 4.1.

M. G. Natrella. *Experimental Statistics.* National Bureau of Standards, Washington, D.C., 1963. National Bureau of Standards Handbook 91.

J. A. Nelder and R. Mead. A simplex algorithm for function minimization. *Computer Journal*, 7:308–313, 1965. doi:10.1093/comjnl/7.4.308.

L. S. Nelson. Control charts. In S. Kotz, N. Balakrishnan, C. B. Read, B. Vidakovic, and N. L. Johnson, editors, *Encyclopedia of Statistical Sciences.* John Wiley & Sons, Hoboken, NJ, second edition, 2005. ISBN 978-0-471-15044-2.

R. G. Newcombe. Two-sided confidence intervals for the single proportion: comparison of seven methods. *Statistics in Medicine*, 17(8):857–872, 1998. doi:10.1002/(sici)1097-0258(19980430)17:8<857::aid-sim777>3.0.co;2-e.

NIST/SEMATECH. *NIST/SEMATECH e-Handbook of Statistical Methods.* National Institute of Standards and Technology, U.S. Department of Commerce, Gaithersburg, Maryland, 2012. doi:10.18434/M32189. URL `https://www.itl.nist.gov/div898/handbook/`.

J. Noh and G. Danuser. Estimation of the fraction of COVID-19 infected people in U.S. states and countries worldwide. *PLoS ONE*, 16:e0246772, 2021. doi:10.1371/journal.pone.0246772.

A. O'Hagan. The Bayesian Approach to Statistics. In T. Rudas, editor, *Handbook of Probability: Theory and Applications*, chapter 6. Sage Publications, Thousand Oaks, CA, 2008. ISBN 978-1-4129-2714-7. doi:10.4135/9781452226620.n6.

A. O'Hagan, C. E. Buck, A. Daneshkhah, J. R. Eiser, P. H. Garthwaite, D. J. Jenkinson, J. E. Oakley, and T. Rakow. *Uncertain Judgements: Eliciting Experts' Probabilities*. Statistics in Practice. John Wiley & Sons, Chichester, England, 2006. ISBN 978-0-470-02999-2.

D. A. Papanastassiou, G. J. Wasserburg, and D. S. Burnett. Rb-Sr ages of lunar rocks from the Sea of Tranquillity. *Earth and Planetary Science Letters*, 8(1):1–19, 1970. doi:10.1016/0012-821X(70)90093-2.

H. Passing and W. Bablok. A New Biometrical Procedure for Testing the Equality of Measurements from Two Different Analytical Methods. Application of linear regression procedures for method comparison studies in Clinical Chemistry, Part I. *Clinical Chemistry and Laboratory Medicine*, 21:709–720, 1983. doi:10.1515/cclm.1983.21.11.709.

D. W. Pesce, J. A. Braatz, M. J. Reid, A. G. Riess, D. Scolnic, J. J. Condon, F. Gao, C. Henkel, C. M. V. Impellizzeri, C. Y. Kuo, and K. Y. Lo. The Megamaser cosmology project. XIII. Combined Hubble constant constraints. *The Astrophysical Journal*, 891(1):L1, February 2020. doi:10.3847/2041-8213/ab75f0.

Planck Collab. et al. Planck 2018 results — VI. Cosmological parameters. *Astronomy & Astrophysics*, 641:A6, 2020. doi:10.1051/0004-6361/201833910.

P. E. Pontius and J. M. Cameron. *Realistic Uncertainties and the Mass Measurement Process — An Illustrated Review*. Number 103 in NBS Monograph Series. National Bureau of Standards, Washington,

DC, 1967. URL http://nvlpubs.nist.gov/nistpubs/Legacy/MONO/nbsmonograph103.pdf.

A. Possolo. *Simple Guide for Evaluating and Expressing the Uncertainty of NIST Measurement Results.* National Institute of Standards and Technology, Gaithersburg, MD, 2015. doi:10.6028/NIST.TN.1900. NIST Technical Note 1900.

A. Possolo. Measurement. In A. B. Forbes, N.-F. Zhang, A. Chunovkina, S. Eichstädt, and F. Pavese, editors, *Advanced Mathematical and Computational Tools in Metrology and Testing: AMCTM XI*, volume 89 of *Series on Advances in Mathematics for Applied Sciences*, pages 273–285. World Scientific Publishing Company, Singapore, 2018. ISBN 978-981-3274-29-7. doi:10.1142/9789813274303_0027.

A. Possolo. *Evaluating, Expressing, and Propagating Measurement Uncertainty for NIST Reference Materials.* National Institute of Standards and Technology, Gaithersburg, MD, 2020. doi:10.6028/NIST.SP.260-202. NIST Special Publication 260-202.

A. Possolo and H. K. Iyer. Concepts and tools for the evaluation of measurement uncertainty. *Review of Scientific Instruments*, 88(1):011301, 2017. doi:10.1063/1.4974274.

A. Possolo, C. Merkatas, and O. Bodnar. Asymmetrical uncertainties. *Metrologia*, 56(4):045009, 2019. doi:10.1088/1681-7575/ab2a8d.

A. Possolo, A. Koepke, D. Newton, and M. R. Winchester. Decision tree for key comparisons. *Journal of Research of the National Institute of Standards and Technology*, 126: 126007, 2021. doi:10.6028/jres.126.007.

V. Poulin, T. L. Smith, T. Karwal, and M. Kamionkowski. Early dark energy can resolve the Hubble tension. *Physical Review Letters*, 122:221301, June 2019. doi:10.1103/ PhysRevLett.122.221301.

J. B. Quinn and G. D. Quinn. A practical and systematic review of Weibull statistics for reporting strengths of dental materials. *Dental Materials*, 26:135–147, 2010. doi:10.1016/j.dental.2009.09.006.

C. Bronk Ramsey. Bayesian analysis of radiocarbon dates. *Radiocarbon*, 51(1):337–360, 2009. doi:10.1017/S0033822200033865.

C. E. Rasmussen and C. K. I. Williams. *Gaussian Processes for Machine Learning*. The MIT Press, Cambridge, MA, 2006. ISBN ISBN 0-262-18253-X.

M. J. Reid, D. W. Pesce, and A. G. Riess. An improved distance to NGC 4258 and its implications for the Hubble Constant. *The Astrophysical Journal*, 886(2):L27, November 2019. doi:10.3847/2041-8213/ab552d.

P. J. Reimer et al. The INTCAL20 northern hemisphere radiocarbon age calibration curve (0–55 cal kBP). *Radiocarbon*, pages 1–33, 2020. doi:10.1017/RDC.2020.41.

P. R. Renne, A. L. Deino, F. J. Hilgen, K. F. Kuiper, D. F. Mark, W. S. Mitchell, L. E. Morgan, R. Mundil, and J. Smit. Time scales of critical events around the Cretaceous-Paleogene boundary. *Science*, 339(6120): 684–687, 2013. doi:10.1126/science.1230492.

A. G. Riess, S. Casertano, W. Yuan, L. M. Macri, and D. Scolnic. Large Magellanic Cloud Cepheid Standards provide a 1% foundation for the determination of the Hubble Constant and stronger evidence for physics beyond ΛCDM. *The Astrophysical Journal*, 876 (1):85, May 2019. doi:10.3847/1538-4357/ab1422.

A. G. Riess, S. Casertano, W. Yuan, J. B. Bowers, L. Macri, J. C. Zinn, and D. Scolnic. Cosmic distances calibrated to 1% precision with Gaia EDR3 parallaxes and Hubble Space Telescope photometry of 75 Milky Way Cepheids confirm tension with ΛCDM. *The Astrophysical Journal Letters*, 908(1):L6, 2021a. doi:10.3847/2041-8213/abdbaf.

A. G. Riess et al. A comprehensive measurement of the local value of the Hubble Constant with 1 km/s/Mpc uncertainty from the Hubble Space Telescope and the SHoES Team, 2021b.

C. Robert and G. Casella. A short history of Markov Chain Monte Carlo: Subjective recollections from incomplete data. *Statistical Science*, 26(1):102–115, 2011. doi:10.1214/10-STS351o.

C. P. Robert and G. Casella. *Introducing Monte Carlo Methods with R*. Springer, New York, NY, 2010. ISBN 978-1-4419-1575-7. doi:10.1007/978-1-4419-1576-4.

P. J. Rousseeuw and C. Croux. Alternatives to the median absolute deviation. *Journal of the American Statistical Association*, 88:1273–1283, December 1993.

M. Rubino, D. M. Etheridge, D. P. Thornton, R. Howden, C. E. Allison, R. J. Francey, R. L. Langenfelds, L. P. Steele, C. M. Trudinger, D. A. Spencer, M. A. J. Curran, T. D. van Ommen, and A. M. Smith. Revised records of atmospheric trace gases CO_2, CH_4, N_2O and $\delta^{13}C$-CO_2 over the last 2000 years from Law Dome, Antarctica. *Earth System Science Data*, 11(2): 473–492, 2019. doi:10.5194/essd-11-473-2019.

R. Ruggles and H. Brodie. An empirical approach to economic intelligence in World War II. *Journal of the American Statistical Association*, 42(237):72–91, 1947. doi:10.2307/2280189.

E. Rutherford, H. Geiger, and H. Bateman. The probability variations in the distribution of α particles. *The London, Edinburgh, and Dublin Philosophical Magazine and Journal of Science*, 20:698–707, 1910. doi:10.1080/14786441008636955.

J. Ryan, Y. Chen, and B. Ratra. Baryon acoustic oscillation, Hubble parameter, and angular size measurement constraints on the Hubble constant, dark energy dynamics, and spatial curvature. *Monthly Notices of the Royal Astronomical Society*, 488(3):3844–3856, July 2019. doi:10.1093/mnras/stz1966.

L. J. Savage. *The Foundations of Statistics*. Dover Publications, New York, New York, 1972.

L. Scrucca. qcc: an R package for quality control charting and statistical process control. *R News*, 4:11–17, 2004. URL `https://cran.r-project.org/doc/Rnews/`.

S. R. Searle, G. Casella, and C. E. McCulloch. *Variance Components*. John Wiley & Sons, Hoboken, NJ, 2006. ISBN 0-470-00959-4.

G. A. F. Seber. *A Matrix Handbook for Statisticians*. John Wiley & Sons, Hoboken, NJ, 2008. ISBN 978-0-471-74869-4.

T. M. Sedgwick, C. A. Collins, I. K. Baldry, and P. A. James. The effects of peculiar velocities in SN Ia environments on the local H_0 measurement. *Monthly Notices of the Royal Astronomical Society*, 500(3):3728–3742, 2021. doi:10.1093/mnras/staa3456.

A. J. Shajib et al. STRIDES: A 3.9 per cent measurement of the Hubble constant from the strong lens system DES J0408-5354, 2019. URL `https://arxiv.org/abs/1910.06306`.

S. S. Shapiro and M. B. Wilk. An analysis of variance test for normality (complete samples). *Biometrika*, 52 (3,4):591–611, 1965. doi:10.2307/2333709.

W. A. Shewhart. *Economic Control of Quality of Manufactured Product*. D. Van Nostrand Company, Princeton, NJ, 1931.

R. Silberzahn and E. L. Uhlmann. Crowdsourced research: Many hands make tight work. *Nature*, 526: 189–191, October 2015. doi:10.1038/526189a.

J. Silver-Greenberg and S. Cowley. 5 Reasons to Question Breath Tests. *The New York Times*, November 3, 2019. Business Section.

G. Simpson. Accuracy and precision of breath-alcohol measurements for a random subject in the postabsorptive state. *Clinical Chemistry*, 33(2):261–268, 1987. doi:10.1093/clinchem/33.2.261.

T. Simpson. A Letter to the Right Honourable George Earl of Macclesfield, President of the Royal Society, on the Advantage of Taking the Mean of a Number of Observations, in Practical Astronomy. *Philosophical Transactions of the Royal Society of London*, 49:82–93, 1755. doi:10.1098/rstl.1755.0020.

A. Sitek and A. M. Celler. Limitations of Poisson statistics in describing radioactive decay. *Physica Medica*, 31: 1105–1107, 2015. doi:10.1016/j.ejmp.2015.08.015.

D. L. Snyder and M. I. Miller. *Random Point Processes in Time and Space*. Springer-Verlag, New York, NY, 2nd edition, 1991. ISBN 978-1-4612-7821-4. doi:10.1007/978-1-4612-3166-0.

J. Sokol. A recharged debate over the speed of the expansion of the universe could lead to new physics. *Science*, March 2017. doi:10.1126/science.aal0877.

J. Soltis, S. Casertano, and A. G. Riess. The parallax of ω Centauri measured from Gaia EDR3 and a direct, geometric calibration of the tip of the Red Giant Branch and the Hubble Constant. *The Astrophysical Journal Letters*, 908(1):L5, 2021. doi:10.3847/2041-8213/abdbad.

Stan Development Team. *Stan Modeling Language — User's Guide and Reference Manual*. Available at `http://mc-stan.org/`, 2016. Stan Version 2.14.0.

Stan Development Team. *Stan User's Guide*. `mc-stan.org`, 2019. Stan Version 2.28.

T. Stockman, G. Monroe, and S. Cordner. Venus is not Earth's closest neighbor. *Physics Today*, 72, March 2019. doi:10.1063/PT.6.3.20190312a.

L. D. Stone, C. Keller, T. L. Kratzke, and J. Strumpfer. Search analysis for the location of the AF447 underwater wreckage. Technical report, Metron Scientific Solutions, Reston, VA, January 2011. Report to Bureau d'Enquêtes et d'Analyses pour la sécurité de l'aviation civile.

Student. The probable error of a mean. *Biometrika*, 6(1): 1–25, March 1908.

Teva Pharmaceuticals USA, Inc. v. *Sandoz, Inc.* 574 U. S. 318 (2015), 2015.

W. L. Tew and G. F. Strouse. *Standard Reference Material 1750: Standard Platinum Resistance Thermometers, 13.8033 K to 429.7485 K*. NIST Special Publication 260-139. National Institute of Standards and Technology, Gaithersburg, MD, November 2001. doi:10.6028/NIST.SP.260-139.

M. Thompson and S. L. R. Ellison. Dark uncertainty. *Accreditation and Quality Assurance*, 16:483–487, October 2011. doi:10.1007/s00769-011-0803-0.

K. W. Thoning, P. P. Tans, and W. D. Komhyr. Atmospheric carbon dioxide at Mauna Loa Observatory: 2. Analysis of the NOAA GMCC data, 1974–1985. *Journal of Geophysical Research: Atmospheres*, 94(D6):8549–8565, 1989. doi:10.1029/JD094iD06p08549.

H. L. Thuillier and R. Smyth. *A Manual of Surveying for India, detailing the mode of operations on the Trigonometrical, Topographical, and Revenue Surveys of India*. Thacker, Spink & Co., Calcutta, India, third edition, 1875.

E. Tiesinga, P. J. Mohr, D. B. Newell, and B. N. Taylor. CODATA recommended values of the fundamental physical constants: 2018. *Reviews of Modern Physics*, 93: 025010, Jun 2021. doi:10.1103/RevModPhys.93.025010.

J. Todd. The prehistory and early history of computation at the U.S. National Bureau of Standards. In S. G. Nash, editor, *A History of Scientific Computing*, ACM Press History Series, pages 251–268. Addison-Wesley, Reading, MA, 1990. Conference on the History of Scientific and Numeric Computation, Princeton, N.J., 1987.

S. Tong, J. Meija, L. Zhou, B. Methven, Z. Mester, and L. Yang. High-precision measurements of the isotopic composition of common lead using MC-ICPMS: Comparison of calibration strategies based on full gravimetric isotope mixture and regression models. *Analytical Chemistry*, 91(6):4164–4171, 2019. doi:10.1021/acs.analchem.9b00020.

J. R. Townsley. BP: Time for a change. *Radiocarbon*, 59(1): 177–178, 2017. doi:10.1017/RDC.2017.2.

J. W. Tukey. *Exploratory Data Analysis*. Addison-Wesley, Reading, MA, 1977. ISBN 0-201-07616-0.

J. W. Tukey. Methodological comments focused on opportunities. In P. R. Monge and J. N. Cappella, editors, *Multivariate Techniques in Human Communication Research*, chapter 16, pages 490–528. Academic Press, London, UK, 1980. ISBN 0-12-504450-X.

J. W. Tukey. Choosing techniques for the analysis of data. In L. V. Jones, editor, *The Collected Works of John Tukey — Philosophy and Principles of Data Analysis: 1965-1986*, volume 4, chapter 24. Wadsworth & Brooks/Cole, Monterey, CA, 1986. Previously unpublished manuscript.

T. van Hoof, F. P. M. Bunnik, J. G. M. Waucomont, W. M. Kürschner, and H. Visscher. Forest re-growth on medieval farmland after the black death pandemic —

implications for atmospheric CO_2 levels. *Palaeogeography, Palaeoclimatology, and Palaeoecology*, 237:396–409, 2006. doi:10.1016/j.palaeo.2005.12.013.

R. N. Varner and R. C. Raybold. *National Bureau of Standards Mass Calibration Computer Software*. NIST Technical Note 1127. National Bureau of Standards, Washington, DC, July 1980. URL `https://nvlpubs.nist.gov/nistpubs/Legacy/TN/nbstechnicalnote1127.pdf`.

A. Vehtari, J. Gabry, Y. Yao, and A. Gelman. loo: Efficient leave-one-out cross-validation and WAIC for Bayesian models, 2019. URL `https://CRAN.R-project.org/package=loo`. R package version 2.2.0.

W. N. Venables and B. D. Ripley. *Modern Applied Statistics with S*. Springer, New York, Fourth edition, 2002. ISBN 0-387-95457-0. URL `www.stats.ox.ac.uk/pub/MASS4`.

K. R. von Hauer. Über die zusammensetzung des kalium-telturbromides und das Äquivalent des tellurs. *Sitzungsberichte der Kaiserlichen Akademie der Wissenschaften. Mathematisch-Naturwissenschaftliche Classe*, 25:139–144, 1857.

M. Wang, W. J. Huang, F. G. Kondev, G. Audi, and S. Naimi. The AME 2020 atomic mass evaluation (II). Tables, graphs, and references. *Chinese Physics C*, 45 (3):030003, 2021. doi:10.1088/1674-1137/abddaf.

P. H. Warren and G. J. Taylor. The Moon. In H. D. Holland and K. K. Turekian, editors, *Treatise on Geochemistry*, volume 2, pages 213–250. Elsevier, Oxford, UK, second edition, 2014.

M. C. Wendl. Pseudonymous fame. *Science*, 351:1406, 2016. doi:10.1126/science.351.6280.1406.

Western Electric. *Statistical Quality Control Handbook*. Western Electric Corporation, Indianapolis, IN, 2nd edition, 1958.

G. H. White. Basics of estimating measurement uncertainty. *The Clinical Biochemist Reviews*, 29 (Supplement 1):S53–S60, 8 2008.

G. H. White, C. A. Campbell, and A. R Horvath. Is this a Critical, Panic, Alarm, Urgent, or Markedly Abnormal Result? *Clinical Chemistry*, 60(12):1569–1570, December 2014. doi:10.1373/clinchem.2014.227645.

R. White. The meaning of measurement in metrology. *Accreditation and Quality Assurance*, 16:31–41, 2011. doi:10.1007/s00769-010-0698-1.

WHO. Preventing unsafe abortion. Evidence Brief WHO/RHR/19.21, World Health Organization, Geneva, Switzerland, 2019.

E. B. Wilson. Probable inference, the law of succession, and statistical inference. *Journal of the American Statistical Association*, 22:209–212, 1927. doi:10.2307/2276774.

K. C. Wong et al. H0LiCOW XIII. A 2.4 % measurement of H_0 from lensed quasars: 5.3σ tension between early and late-Universe probes. arXiv:1907.04869, 2019. URL `https://arxiv.org/abs/1907.04869`.

Y. Yao, A. Vehtari, D. Simpson, and A. Gelman. Using stacking to average Bayesian predictive distributions. *Bayesian Analysis*, 2017. doi:10.1214/17-BA1091.

Y. Ye. *Interior Point Algorithms: Theory and Analysis*. John Wiley & Sons, New York, NY, 1997. ISBN 978-0471174202.

T. W. Yee. The VGAM package for categorical data analysis. *Journal of Statistical Software*, 32(10):1–34, 2010. doi:10.18637/jss.v032.i10. URL `www.jstatsoft.org/v32/i10/`.

G. U. Yule. On the theory of correlation. *Journal of the Royal Statistical Society*, 60(4):812–854, December 1897. doi:10.2307/2979746.

G. U. Yule. *An Introduction to the Theory of Statistics*. Charles Griffin and Company, London, UK, 1911.

S. L. Zabell. *Symmetry and Its Discontents*. Cambridge University Press, New York, NY, 2005. ISBN 978-0-521-44912-0.

S. L. Zabell. On Student's 1908 article "The Probable Error of a Mean". *Journal of the American Statistical Association*, 103(481):1–7, 2008. doi:10.1198/016214508000000030.

H. Zangl, M. Zine-Zine, and K. Hoermaier. Utilization of software tools for uncertainty calculation in measurement science education. *Journal of Physics: Conference Series*, 588:012054, 2015. doi:10.1088/1742-6596/588/1/012054.

H. Zeeb and F. Shannoun, editors. *WHO Handbook on Indoor Radon: A Public Health Perspective*. World Health Organization, Geneva, Switzerland, 2009. ISBN 978-92-4-154767-3.

M. Zelen. Linear estimation and related topics. In J. Todd, editor, *Survey of Numerical Analysis*, chapter 17, pages 558–584. McGraw-Hill, New York, NY, 1962.

X.-K. Zhu, J. Benefield, T. B. Coplen, Z. Gao, and N. E. Holden. Variation of lead isotopic composition and atomic weight in terrestrial materials (IUPAC Technical Report). *Pure and Applied Chemistry*, 93:155–166, 2021. doi:10.1515/pac-2018-0916.

Index

Made in the USA
Monee, IL
04 November 2026

67386139R10144